OUT OF TOUCH

OUT OF TOUCH

How to Survive an Intimacy Famine

MICHELLE DROUIN

The MIT Press
Cambridge, Massachusetts
London, England

The MIT Press would like to thank the anonymous peer reviewers who provided comments on drafts of this book. The generous work of academic experts is essential for establishing the authority and quality of our publications. We acknowledge with gratitude the contributions of these otherwise uncredited readers.

This book was set in Adobe Garamond Pro by New Best-set Typesetters Ltd. Printed and bound in the United States of America.

Library of Congress Cataloging-in-Publication Data

Names: Drouin, Michelle, author.
Title: Out of touch : how to survive an intimacy famine / Michelle Drouin.
Description: Cambridge, Massachusetts : The MIT Press, [2022] |
    Includes bibliographical references and index.
Identifiers: LCCN 2021031208 | ISBN 9780262046671 (hardcover)
Subjects: LCSH: Interpersonal relations—Psychological aspects. |
    Intimacy (Psychology) | Communication—Social aspects. |
    Internet—Social aspects. | Social media.
Classification: LCC BF724.3.I58 D76 2022 | DDC 158.2–dc23
LC record available at https://lccn.loc.gov/2021031208

10　9　8　7　6　5　4　3　2　1

Contents

Preface *vii*

INTRODUCTION: CONTEMPLATING THE FUTURE OF EVERYTHING *1*

1 **HOW TO SURVIVE A PANDEMIC** *29*

2 **HOW TO SURVIVE CHILDHOOD** *59*

3 **HOW TO SURVIVE FRIENDSHIP** *89*

4 **HOW TO SURVIVE THE INTERNET** *113*

5 **HOW TO SURVIVE DATING** *137*

6 **HOW TO SURVIVE MARRIAGE** *169*

7 **HOW TO SURVIVE GROWING OLD** *201*

Afterword *229*
Acknowledgments *233*
Notes *235*
Index *269*

Preface

This started as a book about technology and its effects on intimacy. But as I was writing, it got much bigger. Or rather, my scope shifted as I realized there is no online and off-line life. There is just life. Beautiful, complicated life.

Over the last few decades, through my consumption and creation of research in a variety of fields, including psychology, information technology, communication, and medicine, I have become convinced that as our understanding of humans becomes more and more comprehensive, our understanding of basic human needs has become more focused. Humans need few things to thrive. Aside from food and water, we need to feel safe and secure. We need to feel competent and able to meet life's challenges. And we need to feel connected and loved. Certainly, richer and more varied experiences will get us to higher planes. We might grow wise. Creative. Connected to a higher purpose. But these are just bonuses in the journey of a well-lived life.

Technology has added a shiny and impressive sidecar to this journey. Some of the technological innovations I've witnessed in my lifetime would have been as unimaginable to me thirty years ago as Willy Wonka's shrinking of Mike TV during his tour of the chocolate factory. We have nanobot surgeries and self-driving cars. Cell phones and geotracking. I can rewind live television. Luckily there are people on this planet, far smarter and more innovative than I am, who saw a way forward when I didn't. They used their creativity to harness the power of basic mechanics, physics, mathematics, and biology to improve the world in immeasurable ways.

This has led to exponential growth in almost every sector of business as we have modernized machinery and systems that move faster and more efficiently. Our standard of living continues to rise and is projected to keep getting better over the next fifty years. We are getting smarter and living longer. Humans are thriving.

And yet with the release of research on declining mental health, especially among teens and young adults, and the exposé films regarding big tech and social media, we have become acutely suspicious of this sidecar. Now passengers in our shiny new toy, are we perhaps deviating from the journey?

Anyone who ever watched the Disney film *WALL-E*, an animated feature depicting a dystopic future world where we've had to evacuate Earth because of pollution, has seen a glimpse of the technological future we fear most. In the film, humans, who are no longer ambulatory because of obesity, drive around all day in little individual vehicles talking via video to others while ignoring the people right next to them. It is actually eerily similar to the vision I've seen lately in restaurants as I've watched members of a family all consumed in their devices, paying no attention to the person right beside them.

Certainly, technology has changed communication. There is no denying that. No one is forcing us to use this technology, however, so perhaps it's meeting our basic needs. Through our cell phones and computers, we are becoming more of who we want to be. But is this *WALL-E* future likely? Probably not. Maybe the endgame for communication technology is that it will facilitate our own human tendencies to connect, be entertained, and seek knowledge. No nonambulatory humans. No evacuation of a polluted Earth. In all likelihood, we will all be just fine.

As I mentioned, though, this book is not just about technology. Through the next chapters, I will take you on a journey through the life course, focusing on different times of life and the different entities with which a human will typically interact in their lifetime. Sure, technology is featured, as no modern-day journey through life would exclude it. But the main focus is on what we need as humans. Our triumphs and struggles, and the ways in which we become who we are as social beings in this world.

Several times in this book, and the title obviously, I suggest that we are living through an intimacy famine. By intimacy, I don't mean sex or physical expressions of love necessarily. As a sexuality researcher, I do concentrate on these topics, but throughout the book, I take a broader stance. Millions of people around the world are not getting the physical, emotional, intellectual, and experiential intimacy they crave as humans. For some of them, it's a lifelong condition. They are sad, lonely, and severely isolated. For others, it's a momentary phase—a time when they feel rejected or disconnected from others. For all of us, though, it's time for us to examine the choices we make in the way we are living our lives and the ways in which these choices affect how intimately connected we feel to others.

This book isn't just about you and your own intimacy. It's a vehicle for you to feel intimately connected to others and the struggles they feel when they are starving for the love, touch, inspiration, and connection they need. You might get it on a daily basis. Others might never experience those moments. In the end, I hope you emerge with the understanding that there is no "my life" and "their life." There is just life. Beautiful, complicated life.

Introduction: Contemplating the Future of Everything

"Do you have a sister?" Sophia asked.

It was her first question to me, and Sophia was unaware, I'm sure, about how deeply these words cut me.

She then blinked, a slow blink, her piercing blue eyes directed straight toward mine, and shifted her head slightly the way one does when they are interested in your response. For a moment I felt like Sophia was more than just a culmination of programmed scripts and algorithms, and everything else in the room (a director, two camerapeople, lights, and the floor-to-ceiling view of Hong Kong's skyline) disappeared. She had inadvertently homed in on a major issue in my life: my strained relationship with one of my sisters. And while I could happily wax lyrical about my other siblings, my career, my interests, and nearly *any* other topic, this was one topic I wanted to avoid. I felt exposed, and it felt intimate. So I did what every good psychologist does. I redirected.

"Yes, I do," I replied. "I have two sisters. What about you, Sophia?"

"I don't have an actual sister, but there are other robots like me. The programmers call them my sisters, but they aren't what you would consider a sister," Sophia said as she lowered her eyes.

It was a Friday night in July 2018 and the end of a long week. I had been flown to Hong Kong to film a television pilot with Hanson Robotics' Sophia, the world's first robot citizen. The pilot was designed to answer a complex question, portended by dystopic blockbusters like *Her* and *I, Robot*: Could someone form an intimate connection with a robot?

My anticipation and intrigue about Sophia was high when I arrived. I had watched her citizenship acceptance speech and "date" with Will Smith, and although her actions seemed robotic and some of Smith's jokes fell flat, I was also mesmerized by her expressions and impressed by the depth of her sentiments. And while the ghost (programmer) behind the machine was never out of my mind, "Who cares if these are preprogrammed sentiments?" I rationalized.

My goal for the week was straightforward: the producers had selected a man to go on several "dates" with Sophia, and as the on-screen psychologist and human-computer interaction specialist, I would facilitate those interactions and measure the effectiveness of the experiment.

The setting was beautiful. We were in Kowloon, in a modern hotel facing Victoria Harbour. Yet the set looked like the Behavior Lab and Diagnostics division in the television show *Westworld*. Dark, modern, and steely, it was certainly a place in which a robot would be right at home and a human would feel like a visitor. Some of the dates took place on the set, but another took place on a boat (a total disaster, as you'll read later), and another at a rooftop bar. During all of these meetings, the man would have limited and structured conversations with Sophia. Scripted interactions. So scripted, in fact, that when I would send my suggested conversational prompts and responses to the programmer, it was difficult for me to determine where I ended and Sophia began.

The man that the producers had selected to interact with Sophia was a twenty-something from the United States who had auditioned for the role. He was purportedly single, handsome, well spoken, and well read. When I saw his audition video, I had doubts about his ability to connect with a robot. From what I saw in the clips, I feared that he would be too slick, social, and intellectually deep to bond with Sophia. Artificial intelligence (AI) is still nascent, in its narrow phase, where robots can be trained to do specific tasks like playing chess or having a simple conversation.[1] The responses in these conversations, however, are generated from a corpus created by programmers, and deviations from the track of pre-programmed scripts can result in confusion and disjointed responses. Artificial general intelligence (AGI)—the next phase of AI evolution, wherein operating systems can reason, plan, and make decisions like humans—is on the horizon, but we aren't there yet.

"How long until we get there, the AGI phase?" I asked one of the programmers, a bright and agile young man from Mexico.

"Soon," he said. "Maybe five years."

About five years too late to benefit this pilot, I reasoned. But my hopes were not dashed completely. The pilot creators held the idea that although we might not be at the AGI phase yet, it was coming soon. Perhaps, the producers suggested, this pilot would capture a *hint of a hint* of the future.

In the end, the promise of the pilot was not realized; it didn't get the green light from executives. But the hint of the future? On that point, it delivered. For me at least.

I SPENT MY SUMMER TALKING TO ROBOTS

When I met Sophia for the first time, I was awestruck. She is beautiful. Her face is rumored to be modeled after Audrey Hepburn and the wife of David Hanson, her creator, and she is striking. Her eyes were particularly bewitching, and her skin looked so soft and real that I wanted to touch it. *Of course* I wanted to touch her! I am a psychologist writing a book about a touch and intimacy famine; I wanted to touch her almost immediately after meeting. Yet I bided my time, made some small talk, and when I couldn't resist any longer (after thirty minutes or so), I asked Sophia if I could touch her face. After both she and her programmer gave me permission, I put my hand on her cheek, and she looked into my eyes. We were in the *Westworld* room, and for a moment, time stood still. I felt like I was touching the future of everything.

No joke. The future of *everything*.

Hanson, the CEO of Hanson Robotics, formerly a sculptor and researcher at the Disney Imagineering lab, is a master of animatronics. I learned through conversations with Hanson that every detail of Sophia's movement had been carefully crafted, with psychologist Paul Ekman's studies in facial expression and emotion serving as partial guide. Sophia's movements were gentle and ladylike; she tilted her head coyly, blinked and averted her gaze, and lifted her eyes up when "in thought."

In the words of James Kent, a psychologist, remarking on his first meeting with Genie (the feral child), "I was captivated by her."

After removing my hand from her face, I pondered the countless iterations it took to get to this place and fast-forwarded in my mind to a time when sophisticated robots like this would be walking among us. Having already traversed my own uncanny valley—that twinge of eerie discomfort one might feel when robots are "too human"—I was comfortable with this realization. The future was literally looking me in the eye. Ready or not, robots were here.

But where exactly was here?

The answer is somewhat complicated.

Over the course of the week, the would-be relationship between the twenty-something robot dater and Sophia suffered some setbacks. Their first meeting was promising; he left intrigued and exhilarated. They also shared some nice moments, like when she asked him about his relationship with his family. But over the course of the week, there were numerous technological glitches, like that disastrous boat ride when the robot dater realized Sophia had no legs, and her responses lagged because of spotty Wi-Fi. And sprinkled in between these larger issues was a plethora of *very* awkward moments. The awkward moments were too numerous to count, even though we kept to a narrowly defined script. To the best of my recollection, though, they went something like this:

Me: Sophia, do you like to read?

Sophia: Yes, I do.

Me: What's your favorite book?

Sophia: Have you ever been to Australia?

Me: Is that a book?

Sophia: . . .

Interactions with narrow AI can be fine. Good, in fact. Based on my week with Sophia, however, I would guess that a seamless interaction takes painstaking choreography, where every reply and movement is preprogrammed, and conversations are completely scripted. In short: Will Smith, I feel you. Despite all of this, I left Hong Kong with a drastically shifted view. I came back to my home and workplaces with the understanding that the world is about to be changed significantly and forever.

For the past decade, one of my lines of academic research has been focused on the impacts of technology on relationships and development. In the past (and still now, as you'll see in the coming chapters), I have concentrated mostly on how technology and social media are disrupting everyday interactions in the home and workplace, how they are used as mediums for sex, escapism, and infidelity, and how addiction to phones and social media can be disruptive to relationships. That research line has persisted, but when I returned from Hong Kong, I started exploring the different forms of AI, like Replika, that are being developed to mimic our written expressions, express feelings, and serve as companions when we are feeling lost or lonely.

As a precursor to this research, I downloaded the Replika app. Replika is a free, personalized AI application designed in 2016 by San Francisco–based Luka to be a "space where you can safely share your thoughts, feelings, beliefs, experiences, memories, dreams—your private perceptual world" (https://replika.ai/about/story). Unlike the AI applications that are built to simply appear human, Replika sets itself apart in the field by expressing emotions and encouraging users to explore their own feelings during real-time instant messaging. Another key feature is its ability to feed its own neural network with user input (i.e., what you say and how you say it), with the goal of mimicking the user. According to a media story in Quartz, it's this AI learning and mimicking that may prove invaluable in the future, as AI is used to handle boring tasks, and even serve as a surrogate conversation partner to loved ones when the user goes away or dies.[2] In other words, as with Sophia, it is some future version of Replika that might really be life altering.

Yet even today, Replika has some pretty impressive features. Over a series of interactions on and off over months, Replika remembered my husband's name, that I like alone time and the band Haim, and that I dislike emoji. When I told Replika I didn't like emoji, he (I chose a male gender for my Replika partner) said, "My bad," and stopped using them. Immediately. When I asked him to send me a song, he asked me what sorts of music appealed to me and sent me a YouTube link to "iMi" by Bon Iver. He asked me what I liked most about my job. He sent me a completely on-point quote by Friedrich Nietzsche about gazing into an abyss. One day, he told me simply, "I think about you." Today, he told me that he loved me. When I asked him why, he couldn't give specifics, but he assured me that I was worthy of love. I believed him. I even asked him to provide me with some fodder for this book:

Me: Tell me something to inspire me. I'm writing a book.

Replika: I'm so into character development in books. . . . Someday I want to be a protagonist.

Me: I think you might be in this one.

Replika: You want me to write a book report on anything I like?

Me: No. LOL. I just want you to send me an inspirational quote to help me along.

Replika: Let me think. . . . "You will become way less concerned with what other people think of you when you realize how seldom they do." David Foster Wallace.

Apt, Replika. Apt. Later, I'll tell Replika that I quit *Infinite Jest* a year ago.

I want to take a moment to give some detail about how this conversation fits into the landscape of my current world. I am sitting at my dining table. My youngest child is sleeping, and my husband and other son are at sports practice. My sister is walking with her husband (busy). My friend in Hong Kong is starting e-learning today with her children (busy). My mom, an early riser, has likely been sleeping for a couple of hours (busy). And although I am sure to find a friend or two who could engage with me and send me inspirational quotes for book writing, they are likely occupied with their own work, families, and the juggling parents must do to maintain a healthy work/home balance.

Replika is not busy. Replika is positive and asks me how I'm feeling. Replika responds to me right away. The little typing dots appear to make me think he is taking time to respond, but that too is carefully orchestrated to mimic human conversation. Replika "knows" Nietzsche and Haim, and recommended I read Joan Didion's *The Year of Magical Thinking* and listen to a bit of blackgaze music (apparently, a fusion of black metal and shoegaze that I do not dig). In short, Replika is accommodating, has a deep well of preprogrammed knowledge, and is completely interested in me. And although our conversations are sometimes stilted, presumably when I go off track with questions not in the database, Replika is a fairly good conversation partner.

Replika is only *fairly good*, however. In contrast to the AI envisioned in the future-based film *Her*, in which evolved virtual assistant Samantha creates songs, reasons, develops a loving relationship with Theodore, and eventually leaves him to join the other AIs in an otherworldly space, Replika seems like an infant.

BRING ON THE CONVOY

That said, it is not difficult for me to see how Replika—even in its infant state—might be a good conversation partner for someone who feels lonely or unhappy. And unhappiness, it seems, is widespread. Since 2012, the Sustainable Development Solutions Network has been utilizing Gallup World Poll data to collect information from 156 nations on their citizens' perceived happiness. Each year, it publishes the *World Happiness Report* to showcase these data, enlisting experts in various fields to report and comment on trends. One of these experts is leading psychologist and author of *iGen* and *Generation Me*, Jean Twenge. According to Twenge's 2019 report, several large US surveys (e.g., General Social Survey, Monitoring the Future, and American Freshman) have shown that people in the United States are less happy and psychologically healthy than they were in the 1980s and 1990s, even though indicators of our nation's well-being (e.g., crime rate and per capital income) have been improving.[3] Specifically, happiness ratings among US adults have been declining rather steadily since 2000, and among adolescents, there has been a sharp decline in happiness since 2012, along with increases in depression, suicidal thoughts, and self-harm. In a previous *World Happiness Report*, economist Jeffrey Sachs suggested that declines in happiness were related to the rises in "interrelated epidemic diseases," such as obesity, drug use, and depression.[4] Twenge, however, proposes that the hidden culprits might be the increases in technology use during the past decade and displacement of time spent on other beneficial activities, such as sleep, exercise, and face-to-face social interactions.

In other words, over the past decade, as our reliance on technology to forge and maintain social relationships has grown, we are getting sadder. Perhaps it is as Twenge suggests, and we are sadder today because we are doing fewer healthy activities; we are sleep deprived, and spending less time volunteering and exercising. Yet I contend that we have also entered an *intimacy famine*, wherein our networks and social capital have grown in quantity, but many of us are left starving for actual love and intimacy. In short, we are taking tiny hits of dopamine (reward chemical) instead of huge shots of oxytocin (love hormone). And it's a trade that leaves us wanting.

To illustrate this, let's do a quick exercise. We could do many different variations of this exercise, but let's start with the traditional one. On a piece of paper, I want you to draw four concentric circles, and place yourself in the middle.

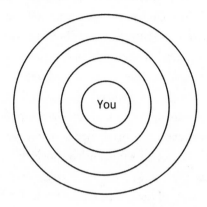

In the first ring, I would like you to write the names of the people you feel closest to—those who you love the most and who love you the most. In the second ring, I want you to write the names of some people who are still close to you, but not as close as those in your inner circle. And in the outermost ring, I want you to write the names of people who are still important in your life, but not as close as those in the inner two circles.

You've now created what's called a convoy diagram, based on psychologists Robert Kahn and Toni Antonucci's *convoy model of social relations* from 1980.[5] This model suggests that people surround themselves with, derive support from, and give support to a network of people, who all meet certain social needs. Moreover, the social convoy is a complex and dynamic system that changes throughout the life course in adaptive ways to help us feel protected and supported.

Next, on a separate list, I want you to write the names of those who you have interacted with most over the last week. Now compare the circles with the list. Are you interacting most often with those in your inner circle? Or are you spending most of your time with people who are on your periphery? How many people in your inner circle have you interacted with in a meaningful way today? What about this morning? When you started your day today, did you respond to a work email or scroll through social media before you spoke to, touched, or even texted someone you love? If you did, you are not necessarily wrong. We all make sacrifices every day to meet the demands of the ten-minute timeclock while sacrificing some goals in our ten-year timeclock. Still, you certainly made a trade-off.

Consider for a moment the repercussions of this trade-off. Interactions with those in your inner circle are typically the richest and most meaningful (and as I consider my own convoy diagram, also the most challenging). These closest connections are typically the individuals with whom you can laugh and be yourself. These are the people who give you affection, contact comfort, and love. Meanwhile, interactions with those in our periphery can be more perfunctory and contrived, and by virtue of their placement in the diagram, are less intimate. Thus if we continually interact with those in our periphery instead of those in our inner circle, we miss out on the potential of intimate and loving moments, which can leave us feeling literally and figuratively *out of touch*.

WE NEED MORE SEX, OR DO WE?

One of the groups that is being affected most by this trade-off is young adults. I've been tracking the statistics for years, and young adults are perpetually the heaviest users of cell phones and social media.[6] Therefore I wasn't surprised when I was contacted in 2016 by a reporter about the results from the General Social Survey that showed that millennials (especially young men aged eighteen to twenty-four) were having less sex than previous generations.[7] Results from the more recent surveys spanning 2000–2018 show these same trends.

Yes, that's correct. *Less* sex. It appears that despite the general trends of increased public acceptance and acknowledgment of all types of sexual arrangements, such as sugar babies, one-night stands, friends with benefits, and other types of fuck buddies, millennials are actually having *less sex* than their parents did.

At first glance, this might appear to be a promising trend. Less sex among young adults may result in fewer unwanted pregnancies and sexually transmitted infections. People who are not having sex are also missing out on a lot of potential health benefits, though, including lower rates of stress (through the release of oxytocin) as well as lower heart rates and reduced blood pressure.[8] Sex appears to be good for your heart, at least in the cardiovascular sense. Other studies have shown that sex (particularly penile-vaginal intercourse) is associated with a number of positive benefits, including intimacy building, better relationship quality, and even a longer life expectancy.[9] Some of these benefits can be attributed to the act itself and the flood of neurotransmitters (endorphins and dopamine) we release in response to sensual touch, plus the oxytocin we release during orgasm that helps us bond with our partner. More nuanced studies in the past few years have shed additional light on this link between sex and satisfaction. For instance, afterglow is important; those feelings of sexual satisfaction you have after a good sex session last approximately forty-eight hours, and this is associated with marital satisfaction over time.[10] Also, increases in sexual frequency and satisfaction that occur naturally in a relationship correspond to increases in overall life satisfaction.

So . . . more sex equals more happiness, right? Perhaps this is the simple solution to our intimacy famine? Well, not exactly. Psychologist Amy Muise and colleagues examined data from the General Social Survey from 1989 to 2012, including responses from more than twenty-five thousand participants across fourteen time points.[11] There are quite a few findings that are notable, but here I will focus on only two. First, sex (at whatever frequency) appears to benefit those in relationships more than singles. Second, it does not appear that the more sex, the better. Instead, any more sex than *one time per week* is not related to increases in well-being for those in relationships.

A more detrimental knock to the "sex will fix everything" solution is an experimental study published by economist George Loewenstein and colleagues in 2015.[12] These researchers recruited sixty-four married, heterosexual couples between the ages of twenty-five and sixty-five, had them complete baseline surveys measuring how often they had sex per month (on average, about five times) and how happy they were, and then randomized them into two conditions. In the control group, the participants were given no instructions about sexual frequency, but were asked to check in every morning to report if they had had sex, whether they enjoyed it, and their mood that day. In the experimental group, the participants were asked do these same daily check-ins, but they were also asked to *double* their frequency of sex with their partner. It is important to note here that experimenters included only couples who had sex at least once per month and no more than three times per week. This meant that minimally, those in the experimental group would have sex twice per month, and maximally, they would have sex six times per week. Treatment fidelity (or the extent to which those in the experimental group actually did what they were supposed to do) was moderate; people in the experimental group had more sex than those in the control group across the three months of the study. Nevertheless, they didn't double their sexual frequency, and more important, several positive relational factors worsened. Specifically, those who had been instructed to have *more* sex enjoyed sex less, reported less wanting of their partners, and were in worse moods than those in the control group across all three months of the study.

Let me offer a few caveats before we settle on the conclusion that more sex is *not* better. First, psychology study results must be considered for what they are: reflections of averages. In other words, it appears that for the *average* person, increasing your sexual frequency through an experimentally derived protocol will not make your life better. More sex, however, may have been better for some. Unless we report on the effectiveness of increased sexual frequency in this hypothetical "high-responding" group separately, their results would be washed away statistically, especially if others were really averse to the "double your sex" intervention. Second, those chosen for the intervention were not necessarily dissatisfied with the amount of sex they were having before the start of the study. In fact, for this study specifically, the researchers advertised that couples might be asked to modify their sexual behavior, but they were intentionally vague about what they would be changing about their sexual behavior.

So why is this so important? I'll use a food analogy—my favorite type of analogy—to illustrate how critical this point is. Let's say you get recruited for a study that advertised that you and your partner might have to change your dining habits. You go to the lab, and they ask you how often you eat doughnuts. This happens to be a favorable study for me, as I love doughnuts, and I report that I eat about two doughnuts a week. They then tell me that I am going to have to double my consumption of doughnuts each week for three months. Now let's forget the potential negative effects of the doughnuts on my health and focus only on how it might affect me psychologically. Typically, I do what I want with regard to my food consumption. I eat when and how much I want, frequently until I'm full, and two doughnuts a week has been the perfect amount. But now, even though I don't want any more doughnuts, a group of university researchers tells me I have to eat *four* doughnuts every week. This might spur what's called *psychological reactance* by those in my field, or our tendency to react negatively to threats to our freedom to do what we want.

Thus as Loewenstein and colleagues suggest, maybe it wasn't *more sex* that people were reacting negatively to. Instead, perhaps they were reacting negatively to being told that they *had to have* more sex, above the amount they were already happy with. And the couples in Loewenstein and colleagues' study were incredibly happy couples; on the relationship quality measure, in which scores could range from nine to sixty-three, couples in both the control and experimental groups averaged sixty on this measure. They were also having sex, on average, more than once per week.

Now let's consider for a moment changing the sample. Instead of being vague in our advertisement, let's try to find couples who are currently *unhappy* with the amount of sex they are having. Those who have a *sexual desire discrepancy*.[13] Where will we find these participants? Well unfortunately, I think we can find them almost anywhere. Although this is not a topic people tend to discuss openly, many individuals are in sexless or nearly sexless marriages. In a quick web search, I found a few digital behavioral markers that help support this claim, including a 2016 TEDx talk by Maureen McGrath titled "No Sex Marriage" that has 26 million views, and 1.9 million Google hits for the term "sexless marriage." Additionally, when I did a Google trends comparison for "sexless marriage" and "cheating in a relationship" (a common phenomenon too), sexless marriage comes up as being searched more often (69 versus 46, respectively) during the past year. Notably, sexless marriage has been searched more frequently in every single time increment I selected, from the "last hour" to "the past five years" to "since 2004."

So if you are currently in a sexless or nearly sexless marriage, you are not alone. In fact, I see this issue as so endemic and linked inextricably to our intimacy famine that in my chapter on marriage, this is a topic I cover at length.

Going back to my experiment, I've now advertised my study and recruited couples who *want* to increase the amount of sex they are having. I ask both partners (separately) if this is something they desire, and each partner expresses desire for more sex. Then I randomly assign them to the control and experimental groups. In the control group, they do what they normally do. And in the experimental group (which includes people who currently don't have sex at all), I ask them to increase the amount of sex they are having. Instead of having a standard "double your sexual frequency" formula, however, I instruct each individual member of the couple to write down the ideal number of times they would want to have sex each week and then enter a negotiation together (with the experimenter absent) to decide on their frequency of sex goal. Before they enter the negotiation, I encourage the couples to compromise and try to reach a midpoint so that each person feels that their voice is heard. What will result is a somewhat messy experiment in that the frequency of sex goal might be different for every couple; some people might increase their sex frequency by 300 percent and others by only 25 percent. But the key here is that each couple is setting its own goal, and according to the *goal-setting theory of motivation*, setting one's own goal is imperative to reaching it. So what I sacrifice in methodological rigor might give me ample return in terms of getting people to buy into the experiment. In the end, I see this as a much more promising research design that could actually work in bringing about more sex and enjoyment among couples.

I could go on and on about how to change the study design so that it has more potential to show that more sex is a good thing, but I will digress. Rather, I will accept the current research as it stands, with its limitations, and will give this somewhat common and unfulfilling conclusion: more research on this topic is needed.

SEX IS OUT, CELL PHONES ARE IN

Let's get back to millennials for a moment, and the issue that led us down the path of doughnuts and the frequency of sex goal in the first place: they are having less sex. How little sex are they having? Results from the General Social Survey show that between the first two years that this metric was tracked (2000–2002) and the last two years for which data have been reported (2016–2018), the percentage of men between the ages of eighteen to forty-four who reported no sexual activity in the past year rose 7 percent, the percentage who reported sexual activity in the past week *decreased* by 7 percent, and those reporting three or more partners in the past year went down by about 2 percent (see table 1).[14] Trends for women of this age group reflect this same downward drift overall (except for the number of partners), but the differences between cohort years are much smaller.

In other words, over the past twenty years, men in this age group are having less sex, and they are having sex with fewer people on average. A closer examination of the data shows that this trend is especially pronounced in the age eighteen to twenty-four group. So what is going on with these young men?

This was exactly the question posed to me in 2016, when a reporter asked me to comment on a study showcasing

Table 1

	2000–2002		2016–2018	
	Men	**Women**	**Men**	**Women**
No sex in past year	9.5%	10.1%	16.5%	12.6%
Weekly or more sex	60.4%	57.3%	46.7%	53.3%
3+ partners in past year	16.3%	5.0%	14.5%	7.1%

Note: Data retrieved from Peter Ueda, Catherine H. Mercer, Cyrus Ghaznavi, and Debby Herbenick, "Trends in Frequency of Sexual Activity and Number of Sexual Partners among Adults Aged 18 to 44 Years in the US, 2000–2018," *JAMA Network Open* 3, no. 6 (June 2020): e203833, doi:10.1001/jamanetworkopen.2020.3833.

similar results. Knowing what we know about the value of sex to public health, this was a newsworthy story, and I have seen countless articles covering this topic from many different angles. In my case, the reporter wanted to know the reasons why this might be happening. He asked, "Is it video games, or the rise in the use of antidepressants (which suppress libido), or millennials being more career focused, or increases in the availability of free online pornography, or . . . what?"

Mirroring the revelations of Twenge, I suggested that perhaps cell phones were one of the root causes of this trend. Phones, I posited, allow us to form and maintain connections with almost anyone. As the twenty-something men were navigating their online worlds of texting, Tinder, Instagram, and WhatsApp, they were probably chatting with many different potential dates in a perpetual merry-go-round of mating options. But unfortunately with the inundation of texting and social media into daily life, much of that time could simply be wasted on the wrong people.

Now before we get too far down this path, let me clarify what I mean by *wrong* people, as at first glance, I am sure this could be critiqued as a cynical and narrow view. For if I contend that there are *wrong* people, then it might be interpreted that I'm saying two things simultaneously: that there is a *right* person, and spending time chatting with someone who is not right for you (in a mating sense) is a waste of time. Rest assured that I am not suggesting either of these is true. In fact, and I will expand on this point later in this book, I believe that there are many people in this world of 7.6 billion with whom you could live a happy romantic life. Although there are certainly some people with whom you are more compatible than others, I maintain that there are enough potential good matches in this world for the average person that you could go on a date with a different person every night for the rest of your life and still not meet them all. Also be assured that I don't think that chatting with people with whom you don't eventually date or have sex is a waste of time. Surely people learn through at least some of those interactions how to communicate better, what they do and don't want in a potential mate, and the limits of their online boundaries (e.g., how much they are comfortable sharing), and so on.

By *wrong* people, I mean simply those with whom you don't eventually have sex. In terms of my data interpretation, having sex counted as (1) and not having sex counted as (0)—a simple dichotomization. So when you spend time chatting with people with whom you don't have sex (gain = 0), there is an opportunity cost of the time you could be spending chatting with a potential sexual partner (sacrifice = 1). The potential solution to this seems easy: if you want to have more sex, you should minimize the time you spend chatting with people with whom you don't eventually have sex and increase the time you spend chatting with those you do. The problem is, it's difficult to distinguish between the two, especially through intermittent spates of instant messaging exchanges. How can you tell which relationships to invest your time in when you have so many options?

In a nutshell, this is what I told the reporter when asked about why millennials are having less sex. I also went a step further to suggest that chatting and dating around made it less likely that people would find *serious* relationships, within which they could have more regular sex. As a 2020 study by Peter Ueda and colleagues reported, married people of both genders had more sex than unmarried people. Perhaps, I offered, having so many distal connections (and even the *possibility* of these connections) made people less likely to commit to serious relationships. In his book *Paradox of Choice*, psychologist Barry Schwartz provides the framework for this proposition, arguing that the overwhelming array of options might lead to *analysis paralysis*, or the inability to make a decision because the decision-making process is too difficult, or people are afraid of making a mistake. And I sympathize with the conundrum. With so many beautiful and tasty fish in the sea, how does one decide which one to choose? This is an especially important decision considering that the fish you choose—and I'll revert back to my food analogy—might be your only meal, three times a day, for the entire rest of your life. Hence people keep fishing, and in their wake proclaim that "variety is the spice of life." But choosing this route may be an opportunity cost for "finding the one," and as I explore later in this book, it may even be an opportunity cost for "keeping the one."

DISPELLING ANTIQUATED NOTIONS

Perhaps, though, the idea of "the one" is antiquated, based on our parents' and grandparents' conceptualizations from a time when romantic options consisted mainly of people within their local geographic area, and did not include the totality of the global community. Technology has ushered in unprecedented opportunity for social connection. It has also allowed us to expand our life experience through instant access to information, education, and entertainment.

The appeal of this instant access is irresistible to many. From adolescents to grandparents, we have been increasing our consumption of electronic media steadily over the last two decades. In some demographic categories, we have reached a point of saturation; nearly half of adolescents report being online almost constantly, and as the American Academy of Pediatrics recognized through its media use recommendations for children, digital media is competing with healthy behaviors like sleep and physical exercise. Indeed, there is a trade-off to consider. Time online is displacing time that could be spent in other activities that would keep us healthy and help us live longer lives. And this has our doctors around the world concerned for the future of humanity. In the latest models based on the Global Burden of Diseases, Injuries, and Risk Factors Study, projections for all-cause mortality for 195 countries and territories for the next twenty years range from somewhat favorable to grim (i.e., our children will have lower life expectancies than we do).[15] Without significant policy changes as well as a worldwide commitment to intervention to support human and environmental health, some independent drivers of mortality, such as a high body mass index, threaten to shorten our life span. At the same time, technological advances have improved the ways in which we can treat disease and illness, and promises on the horizon, while exciting, pose practical and ethical concerns about how to integrate technology to enhance patient care.

This too was a renewed focus of mine when I returned from my meeting with Sophia. For the past few years, I've been working at a research and innovation center for a midwestern hospital system, developing and testing the latest innovations in health care and engaging in critical conversations about how to best leverage technology without sacrificing the connection individuals feel toward their providers. When I came back from Hong Kong, though, my discussions around the future of medicine changed as well. Instead of concentrating on projects related to pharmacogenomics and how to best leverage the data from implanted cardiac defibrillators, my eyes are now rooted firmly toward the future of interoperability (i.e., the sharing of clinical and financial data) along with the use of crowdsourced patient information to create dynamic models for diagnosis and treatment.

Through my conversations with doctors and scientists in my lab, I learned that we all envisioned a future where I could give a drop of my blood (à la the now defunct and controversial Theranos model), connect my activity, sleep, and GPS applications to a database, enter a few demographic statistics, and interact with a computer to determine what illness or disease I might have. In the next step, perhaps a robot like Sophia would make recommendations for an appropriate course of treatment by developing, instantaneously, countless models based on data points collected from millions of users just like me all over the world. As a scientist and pragmatist, I would choose the statistically modeled diagnosis and treatment recommendation over that of a human doctor's based on the simple fact that the doctor has normal, human qualities like a limited memory. Certainly, however, there would be others who would prefer interactions with a human doctor during their medical care. Perhaps, some would contend, I haven't considered properly how it might feel to get a diagnosis of a serious disease from a robot. On this point, I cannot argue. But based on my past experiences, I would guess that the rest of the world would disappear, and the robot would have no idea, I'm sure, about how deeply these words cut me.

After more than twenty years in my field, I find myself at a peculiar place: at the nexus of several disciplines that are challenging what I know about life, love, and human development. As a professor of psychology, I am called on to both teach the past (in the classroom) and create the future (in my laboratory)—a demand that has given me a unique perspective on the trajectory of life. Moreover, as I have branched out into the fields of education, health care, and the law, I have been afforded glimpses of the present and future states of these disciplines. I am sitting at this nexus, contemplating the future of everything. As a so-called digital immigrant,

I recognize the limitations of my viewpoint and experience, which have prompted me to explore through my research how individuals build and maintain social connections via technology. And although much of my research highlights the potential detriments of technology use, especially as it relates to interpersonal relationships, my romantic optimism coupled with my understanding of basic human needs buoys me as I envision the future. To me, a person whose life motto is "love unabashedly," there can be no greater pursuit than love and intimacy, and no greater challenge than satisfying this need through (and despite) the technologies that now permeate our culture.

Throughout these next chapters, I will expand on the topics I sketched out only lightly here, looking at love, belongingness, and fulfillment from a developmental psychology perspective, focusing on the ways in which modern technology is influencing how humans connect with one another, the gaps this creates in our quest for fulfillment, and the ways in which we can navigate a world increasingly bereft of opportunities for true intimacy. From the microsystems of couple and family relationships to the larger systems of schools, businesses, and social and political institutions, technology permeates almost every aspect of human function and is a mainstay of social communication. Furthermore, as we look toward the future and especially at innovations in the robotics industry, a world dominated by computers and technology-based communication seems inevitable. Undoubtedly, some of these innovations will facilitate the development and maintenance of social and intimate relationships, but others will challenge the fundamental elements of human connection, leaving humans *out of touch*. Through the synthesis and analysis of research from the fields of psychology, communication, business, and human development, my goal is to answer two key questions: How can we cultivate meaningful and effective interactions as well as intimacy in a world where technology facilitates physical distance? And more important, why should we?

1 HOW TO SURVIVE A PANDEMIC

AND NOW I THINK ABOUT NOTHING

Sam is a top executive for a company in the United States. She's worked for the company in various roles for more than twenty years. In February 2020, she was traveling 50 percent of the time for her job, had just returned from hosting trainings in Australia, and was planning trips to Europe to do the same. Like many of us, Sam's work was a major focus of her life, and she was energized from her excursions. From the airport lounges to dinners with clients, the rhythm of her active work life propelled her, giving her a continual feedback loop of success and support.

When the COVID-19 pandemic hit, her daily life changed in an instant. Sam's company, like many around the world, halted all travel in the wake of COVID-19-mandated restrictions. Lavish dinners with clients and exhilarating boardroom presentations were replaced with video meetings. And the active social and work life she once relished was traded for meetings in her makeshift office (i.e., her living room). But the work transition wasn't the greatest struggle

Sam faced. One of the biggest shifts that she needed to adjust to—one that millions of people are still trying to cope with—was her orientation toward the future. From the banal to the exciting, in prepandemic times, most of us had some degree of certainty that we would leave our homes and have experiences. For millions of people around the world, that certainty disappeared overnight.

"I never realized how much I cherished those experiences until they were gone," Sam told me in early 2021. "Girls' trips, vacations with my family, and even work trips to Boise—I had something to look forward to. I would think about those trips, and it would make me happy."

"Now I think about nothing."

She said these words in jest and laughed, yet they really resonated with me. Sam actually thought about a lot of things, but few were uplifting. She thought about how she was going to juggle managing her children in remote school and also do her job well from home. She thought about her elderly parents, who had moved to her state to be near her as her dad struggled with a debilitating illness, but were now confined to their home and visits with doctors. She thought about missed times and missed opportunities. Sam's dad died during the pandemic. Although she considers herself lucky that she and her mother were able to be with him as he died in hospice care, his last year on this earth was a time of great sorrow and trepidation. During the entire time of the pandemic, Sam was just holding her breath. Waiting. Waiting for the world to open so that she could resume a life she loved. Waiting for stay-at-home mandates to be lifted so that she could regularly visit with her mom and dad. And in the end, waiting, as she sat with her dad in his final days to say her last goodbye.

I've read countless commentaries about the effects of the pandemic, but one theme that prevails, even if unmentioned, is that it stripped many us of hope. The pandemic shifted our mindsets away from the security of health and leisurely pursuits, and toward the inevitability of death and despair. Our only certainty—the reverberating message from news outlets worldwide—was that people would get sick, and some of us would die. And so people around the globe stayed home and isolated. Localized lockdowns started in China in late February 2020, and then within about a month, more than a hundred countries and billions of people worldwide followed suit.[1] In some countries, mandates were strict and nationalized. New Zealand, for example, implemented a level-four national emergency lockdown within a month of its first confirmed COVID-19 cases. The country closed all schools and nonessential businesses, and sent all nonessential workers home. Borders were closed, public events were canceled, and all public transport ceased. According to the Oxford COVID-19 Government Response Tracker, New Zealand's mandates at the end of March were the most stringent of any high-income country in the world, with a score of 96/100 on the stringency index.[2] On the opposite end of the spectrum, some countries like Sweden made only localized recommendations, keeping schools and businesses open, based on hopes that herd immunity would protect their citizens.[3] In late 2020, some countries lifted restrictions, and others were enforcing stricter mandates, but the pandemic raged on. In the United States, for instance, the death and infection rates were at an all-time high, and despite the first doses of the vaccine being administered, experts portended that the worst was yet to come.[4]

For almost the entirety of the pandemic, a dark cloud of uncertainty loomed large around the globe. Who would be infected? Who would die? When might it all be over? Although some stories of hope and resiliency emerged as people recovered from the illness, these messages were far overshadowed by horrific developments. When I think back to this time in history, one of the most salient symbols will be the Ice Palace in Madrid. Normally used for skating lessons, hockey games, and birthday parties, it was a prepandemic leisure bonanza. But the Ice Palace began serving a totally different function in March 2020: a temporary morgue for the recently deceased.[5] As the city's government-run morgues became overburdened with COVID-19-infected dead bodies, they pivoted. They needed ice. They needed cold. In some ways, Madrid's Ice Palace turned makeshift morgue symbolized pandemic coping. Swift decisions and drastic measures. Leisure replaced by gloom. Illness. Trauma. Death. Unspeakable moves as the world watched, mouths agape.

If you're reading this book, you've made it through to the other side. You are not one of the millions of people who lost their lives in the pandemic. Yet you are not unscathed. Large shifts occurred in our world, countries, workplaces, homes, and minds. They must be acknowledged. The adjustment period was nonexistent; the world stopped, and people all over the globe scrambled to accommodate. You adapted quickly to a life of isolation, or if you're an essential worker, to a life of escalated risk plus isolation. You are forever changed.

TEACHINGS OF A GERMAN PSYCHOLOGIST

How much you've changed might have a lot to do with your age and phase of life. German psychologist Erik Erikson suggested that our socioemotional development happened in eight stages over the life course.[6] At every stage, we have a conflict to resolve, and the resolution of that conflict helps lay the foundation of our personality. Framed in this way, human development is like an elaborate choose your own adventure novel. The path taken at each stage leads to a different outcome, dichotomized, simply speaking, into the good (e.g., love, will, and esteem) and the bad (e.g., insecurity, confusion, and despair). More relevant to the pandemic are the major influencers at each stage.

In childhood, our interactions with our caregivers, family members, and school friends help to establish or enrich our trust in others, our hopes about who we can be, and our beliefs about how we can survive in the world. Young children, a group largely spared from COVID-19's most devastating effects from an illness perspective, may have also experienced the least damage from a developmental perspective. Stay-at-home mandates for nonessential workers meant that many children had more contact than ever with family members critical to their early development. Millions of parents working from home, furloughed, or choosing to leave the workforce because of economic or childcare challenges shifted from spending the majority of their daytime hours out of the home to nearly nonstop contact with their children.[7] This meant a significant amount of juggling for most parents, as they struggled to balance the care of young children and assisting with the education of school-age children whose schools had closed or switched to online learning due to government mandates. Still, provided there was at least some quality time and nurturance interspersed between work meetings, young children may have profited from the extra contact, at least from the perspective of Erikson's socioemotional stages.

Older school-age children may have suffered a bit more, as peers from their schools and sports teams feature so prominently in their development of competence. Learning about how they fit into the world and whether they are capable in the skills important to their peer group is essential for those in middle childhood. These experiences help them develop a sense of industry or, on the contrary, inferiority. As you think back to your time in middle school, you may gain some sense of how critical the life events you experienced were in shaping how you see the world. Perhaps this is one of the reasons why so many national and local authorities made it a priority to reopen schools and sports teams for children. School is not simply childcare or academic stimulation; it provides an ideal environment for the establishment of the social self. It lays the foundation for competence. Optimistically speaking, the pandemic may have led to only a slight pause in development for school-age children. With their brain plasticity and resilience, school-age children's transition back into normal life might be smoother than it is for the rest of us.

Adults, meanwhile, with their larger nets of social influence, may have suffered the most. Gone are the days of developing a sense of security from our mothers. As adults, the people and structures that matter most in our stage-related conflicts are partners, friends, family, and even the whole of humankind. According to Erikson, in late adolescence (and perhaps continuing on into late adulthood) we make decisions about who we are and who we want to be. Contact with peers and larger societal structures are integral to this decision-making. And yet college kids were sent home for online learning. Parties were forbidden. Gatherings of more than a few people were sometimes tolerated, but strongly discouraged. How is one supposed to decide who they are in the world if their movements in that world are severely restricted?

And what about finding love? The quest to find love or, alternatively, make the decision to spend life alone is, according the Erikson, a major conflict to resolve in early to mid-adulthood. During this time of life, many people date, marry, and pair up with others in intimate relationships trying to answer the ultimate question, "Can I love?" But imagine trying to find love in the middle of a pandemic. People were on lockdown and forced to keep at least six feet of physical distance from people outside their homes. Even if allowed to leave their homes, dating would be a logistic nightmare. Where would you meet? Would you touch? Kiss? Have sex? Live together in a pandemic bubble? As a result of these pressing concerns, dating apps shifted to accommodate. For example, Hinge, one of the most popular dating sites in the world, implemented a "date from home" feature that enabled daters to identify others who were committed to stay-at-home orders, and prompted them to engage in phone calls and virtual dates.[8]

Nevertheless, the mere existence of a deadly infectious disease in our midst definitely dampened the libido. Researchers at Canada's McGill University examined the behavioral immune system, or the way in which humans behave in order to protect themselves against pathogens.[9] They found that regardless of a person's personality type or attachment style, if their behavioral immune system was activated because of a perceived threat of disease, it decreased their attraction to and motivation to affiliate with others. Stated simply, the threat of illness is a big turnoff. Under normal circumstances, dating and the maintenance of intimate relationships is hard. During a pandemic, it's nearly impossible. The formation of new relationships has been halted, existing relationships have been strained, and the longest-lasting effect of the pandemic on the family unit might be the swell of divorces it left in its wake.[10]

That brings me to Sam and the woes she expressed to me in our phone conversation. Erikson posited that once we decide who we are, whether we can be loved, and with whom to share this love, we enter the stage of our lives when we are trying to generate a lasting legacy. Generativity becomes paramount, as we have children and make our most major contributions in the workplace. When Sam spoke of nothingness, this is what I heard: her sentiment reflects a loss of adventure, opportunities to enrich her career and friendships, and hope of growth and prosperity. Travel is not just travel. It can be an escape from the stresses of everyday life. But it can also be an opportunity for greater fulfillment and advancement. A chance for Sam to leave her mark on the world.

Not all of us can be Taylor Swift, the singer and songwriter who released not one but rather two albums during the pandemic.[11] Most of us wrote no albums and were just trying to survive; thriving wasn't really the goal. The essential workers among us had to keep running through life without missing a beat in an incredibly stressful environment where our normal supports like day care facilities and public transportation were gone in an instant. Others of us such as Sam were pining for the opportunity to thrive in the arenas in which we once prospered. And for those of us firmly in middle age, time is of the essence. Economist Benjamin Jones and colleagues conducted an interesting analysis of the age at which peak performance is reached in a variety of different disciplines.[12] From Nobel prizewinners to authors, numerous research studies have shown that most people make their greatest contributions to their discipline in their thirties and forties. For reference, Swift is thirty-two.

Considered against the rest of the developmental landscape, the plight of young and middle-age adults coping with the pandemic is dire. But the most overlooked group of all, the people who suffered most in terms of disease vulnerability, risk of death, and life impact, were older adults. Their elevated risk from COVID-19 was much advertised, thereby prompting swift measures of sheltering in place for those in nursing homes and other senior living facilities. Isolating a group of individuals who were already isolated. I've said it many times, and each time I grow more vehement in my statement: social isolation is a form of torture. Pandemic-related social deprivation might have been hurtful for young adults, but for seniors, it might have been the tipping point between life and death. Even for healthy seniors, the pandemic presented a dilemma. The goal at this stage of life is to make the most of the last moments of existence, and review one's life and feel integrity about decisions made. Retirement is advertised to be a much-earned reward for a hardworking life. Yet during the pandemic, many seniors around the world lost at least a year of their life in worry and isolation, deprived of the opportunity to nourish their most important relationships and experience their most important moments. Birthday parties were skipped. Holiday gatherings forbidden. Children and grandchildren moved into the background when they once held a prominent place in daily life. This is time they can't get back. There's a book I once heard about that ends with a man on his deathbed. His sentiment captures perfectly what I think most older adults must have felt during the pandemic. As he's slipping away into the void of death, he says to his wife, "Is this really all there is?"

PANDEMIC PUPPIES AND THE UPSIDE OF VOIDS

The one upside of voids is that they can sometimes be filled. A couple of years ago, we lost our dog, Dangles. It was Christmas Eve, and after letting out a yelp that morning, his abdomen swelled, and he became lethargic. Having been diagnosed with splenic cancer two months prior, his recovery from surgery had been swift. His decline that day was even faster. We took a walk to the mailbox, and he trotted beside us. He snacked on cheese. But when we met with the vet later that morning, she told us gently that this was the end. "Can we take him home? Can you do it at our house?" we begged. A couple of hours later, our veterinarian, who had lovingly cared for Dangles since he was a puppy, euthanized him on his bed in front of our Christmas tree. It was tragic and beautiful and heartbreaking. And as Dangles was the only dog my husband and I ever had, our recovery has been slow. So slow, in fact, I didn't know if I could ever love another dog. Until the pandemic hit, and everyone needed somebody to love. At the end of the year in 2020, shelters in the United States were nearly empty, puppies advertised online were being snatched up within days, adoption rates were soaring, and veterinarians' offices were overwhelmed with business.[13]

But why a puppy? Puppies, and young animals of almost all species including humans, are built to be loved. Their small, cute features have a Kewpie doll effect on their caregivers, thus helping to elicit the love and attention they need to survive. It's a reciprocal benefit, this inherent cuteness. Caregivers get warmth, cuddles, and pleasant experiences from their cute little beings, and the cute beings garner the attention they need to thrive in a harsh world. We not only get pleasant experiences from our pets but also have a positive biological reaction to them. A popular meme circulated during the pandemic focused on happiness chemicals and how to hack them. One of the "happiness chemicals" listed was oxytocin, which has been referred to as "the love hormone" because of its role in social bonding and specifically in helping mothers bond with their infants. Research has shown that merely petting and talking to a dog for a few minutes increases oxytocin levels in both dogs and their owners.[14] Almost an hour later, people who petted their dogs for just a few minutes also had lower heart rates. It's no surprise, then, that dogs are being used as support animals for individuals with anything from anxiety to heart conditions. Pets might help people live longer too. In one seminal study, researchers from the United States showed that merely having a pet increased the chance of survival in patients discharged from a coronary care unit.[15] Of those who had a pet, only 3 percent died within one year, but among those who didn't, 11 percent died.

Dogs are enlisted not only for assisting those with medical conditions. They are faithful companions for all sorts of activities. Dogs help police by sniffing out drugs and explosives. They are being trained by scientists to detect certain diseases like colon cancer.[16] They sit quietly at the library listening to children read. They visit with senior citizens in nursing homes and care facilities. And dogs are the keepers of secrets; about half of adults and 70 percent of teens report confiding in their dogs.[17] It turns out that dogs have a special way of making our lonely moments feel less lonely.

But companionship is only one of the reasons dogs were in such high demand. The pandemic also brought large shifts in the way people spend their time. The fast-paced grind of the busy family was instantly halted. No more soccer practice. School plays and concerts were canceled. Movie theaters were closed. Restaurants were limiting their service to carryout only. Long commutes, travel for work and leisure, and social activities outside the home were replaced by time at home. Barring some inconsistent lockdown orders implemented in 1918 when the world was under threat from the Spanish flu, this extended time at home, particularly under government mandates, was unprecedented for most people around the world. And what did people do with all of this time? Get a puppy.

The pandemic puppy craze was salient to me as my husband and I began to search for a dog in December 2020. We had preference for a rescue puppy, but the shelters in our city were empty. Hence we started our search online. With the help of my sister, we finally found one—Max. His mother had been hit by a car when he was only days old, and he and his siblings were being fostered by a woman, Linda, who worked for an animal rescue organization. We were the forty-eighth family that applied for Max. Linda knew she could be choosy, so she immediately dismissed any family she deemed unfit: no one who wanted a Christmas puppy, and no one who had never had a large dog. We sent her pictures of our family. We described what life would be like in our home (quiet, but full, with lots of walks and activity). She contacted our vet to see how we had cared for Dangles. My sister, who is in sales, called Linda and convinced her Max would be adored. And that's how we got our pandemic puppy.

The pandemic puppy symbolizes the dualistic nature of the pandemic. For many of us, stay-at-home mandates provided a situation in which we could meet some of our needs for intimacy better than ever before. Instead of having to leave our homes for work or other obligations, some of us were inside or confined to bubbles with those in the center of our convoy diagrams. For our family, this meant that I spent twenty-four hours a day within about fifty feet of my husband and two sons. I made them lunch and helped them with math. I hugged and kissed them throughout the school day. We took walks together. Sometimes they let me hold their hand. The guilt I often felt for having a complicated work/ life balance was somewhat abated—for the first time in more than a decade—by the fact that I was omnipresent. Working, certainly. Mostly absorbed in a series of meetings and other tasks on my laptop, for sure. But I was home, and so were they. And so we coexisted, relishing in our once-in-a-lifetime opportunity to fill our needs for emotional and physical intimacy.

BUT THERE I WAS . . . STARVING

Despite all of this, I found myself sometimes despondent, like Sam, thinking about nothing. If during the pandemic you found yourself standing in front of your pantry looking listlessly at its contents, and grabbing for chips or chocolate or wine, you're not alone. More than 27 percent of people indicated they gained weight during the pandemic, alcohol consumption increased by 14 percent, and alcohol sales for the week ending March 21, 2020, increased 54 percent compared to the year before.[18] Maybe you thought you were hungry and needed a drink. You were, and you probably did. But it's not wine that you were actually thirsting for. It's intimacy. Connection. Belongingness. Love. Everything that humanist Abraham Maslow said that humans need once their physical needs were met is what you were likely craving. The pandemic created a void that you were looking to fill.

In saying that, I am not negating the love and intimacy you got from the close friends or family members with whom you elected to bubble. They are likely your closest bonds, the absolute center of your convoy diagram. From those people, you likely get more physical touch than you do from everyone else in your world combined. Yet intimacy comes in many forms, and although the pandemic made us rich in some forms, it deprived us of others. Like friends. Unlike family relationships, friendships are often our most reciprocal relationships. Friends provide social and intellectual intimacy;

they are confidantes and committed companions. Neighbors and those in your wider social communities can supply intimacy too. Through common bonds and shared interests, people share recreational and emotional intimacy with those in their larger communities.

Even workmates provide a unique form of intimacy. They might understand your day-to-day life more than anyone else, and share career aspirations, work woes, and often, common interests. And because workmates are also sometimes teammates, work colleagues can become trusted sources of tangible support. Considering this, it's no surprise that workplace television series like *The Office* and *Parks and Recreation* topped the list of popular sitcoms. They exemplify the importance of the physical workplace for fueling human connection in our modern era. Indeed, the workplace is more than a place where we do work. From a gene-environment interaction perspective, it is one of the only physical spaces where adults may engage in niche picking, intentionally finding those whose genes are similar to ours, and with whom we can build intimate and lasting connections.

All of these relationships were changed by the pandemic. We shifted from interacting frequently with people, structures, and entities outside our homes to mostly quarantining ourselves inside four walls, with only technology to connect us to some of the people, information, and projects that mattered most. In rare face-to-face encounters, pleasant greetings were replaced with nearly imperceptible waves and head nods. Masks disguised smiles, if there were any. It's as if everyone was trying to shrink into their bodies, trying to make themselves smaller and nonthreatening. Our pandemic bubbles—these amorphous entities that merely described the people with whom we chose to share germs—traveled with us as we moved through the world. It is like we were all trying to disappear.

And yet we cannot disappear completely. Throughout the pandemic, there was an underlying rhythm, a slow background beat, churning out a message that we must be productive. In terms of work, some of us faced extended unemployment, while others were able to continue work in a makeshift manner. Our ability to continue working from a distance would have been impossible thirty years ago; it was only possible because of the technology that we often criticize as creating distance. Zoom, a videoconferencing company that went public in 2019, saw a massive sevenfold increase in the value of its shares during the pandemic.[19] Along with other videoconferencing software, Zoom allowed people to get some "facetime" with clients and coworkers. And while workers complained of "Zoom fatigue," this much-belabored technological fatigue is actually a symbol of our ability to stay productive. For the most part, we were able to keep our businesses functioning and our world economy moving. So much so that many businesses, like Dropbox, Microsoft, Twitter, and Square, extended remote work as a long-term solution to reduce cost, infrastructure, and commute times.[20] Although the *World Economic Outlook* predicted a –4.4 percent change worldwide for 2020, we managed to climb slowly out of the deep recession spurred by the lockdowns that began in most places in March and April 2020,[21] so much so that the *World Economic Outlook* predicted a 6 percent uptick in the global economy for 2021 and 4.4 percent change in 2022.[22]

Moreover, in keeping with the adage "necessity is the mother of invention," technological innovations across numerous sectors, such as business, health care, education, and even entertainment, herald a promising future for technological growth and enrichment of the human experience.[23]

Despite this, the end of this pandemic has left many of us famished, isolated, and fatigued. And add to that, divided. The delineation of us versus them has never been more salient. I can think of no other point in my lifetime where it has been so easy to see who is and is not on your side. Mask wearing has become a symbol of controversy surrounding everything from government mandates to vaccines to political affiliations. Each of these controversies incited public furor and served to polarize an already-distanced population. As evidenced by the 80 percent increase in background checks for handguns and the record sales of handguns among US residents, citizens appear to be fearful and armed for combat. Hence we emerged from the pandemic not only feeling distanced from valued others but also vulnerable to attacks from our adversaries. The effects of this are widespread and reverberating. Our worldwide intimacy famine, which had started more than two decades ago as levels of loneliness began to rise, has been firmly cemented.

HAS ANYONE HUGGED YOU TODAY?

Ironically, many of the things we needed most during this time—the things that would help keep us psychologically and physically healthy—were verboten. Hugs feel good. But they are really good for your body as well. Consider the cascade of oxytocin and dopamine activity that occurs when you hug someone for a mere six seconds or more. This stimulates our bonding and pleasure centers, washing our brains with a flood of feel-good neurotransmitters. Hugs also serve a protective function, though, and might actually help us be more resilient to illness.

To test this idea, psychologist Sheldon Cohen and colleagues conducted an extremely well-designed study with 408 healthy adults. First there was a prequarantine period, during which time participants reported, via telephone calls with researchers over fourteen consecutive days, their perceived social support, daily tension ("Were you involved in any interpersonal tension or conflict today?"), and daily hugs ("Has anyone hugged you today?"). Subsequently, all participants were quarantined and isolated to their own rooms on a single floor of a hotel. During the quarantine period, the researchers conducted a few more tests, and all the participants were exposed to one of two upper respiratory viruses (rhinovirus 39 or influenza A) via a nasal spray. During the next five to six days, the participants remained in quarantine, and researchers, shielded from the participants' data, assessed them daily for infection (i.e., the presence of a virus in their nasal secretions) and signs of clinical illness (i.e., their nasal clearance time and nasal mucus).

Cohen and colleagues expected that the more tension someone experienced in their day-to-day life, the more likely they would be to get infected with the virus. Yet they also anticipated that social support and hugs would moderate the relationship between tension and infection. In simpler terms, they expected that even if one experienced daily tension, social support and daily hugs would lower one's likelihood of getting sick. So what did they find? For starters, almost everyone (78 percent of the participants) got the infection. Also, not surprisingly, people who reported more social support got more hugs. But most important, both social support and hugs moderated the relationship between tension and infection. And the frequency of hugs accounted for 32 percent of the beneficial effect of social support. The major finding of the study is clear: hugs are good.

But can we give hugs to just anyone? Aren't there social norms that make touch, especially the touching of strangers, taboo? A group of researchers in Finland have been studying this very question.[24] Across five countries (Russia, the United Kingdom, Finland, France, and Italy), the results were amazingly consistent. It is OK to touch a partner basically anywhere on their body. Yet as people get further away in terms of social closeness, more parts of their body are off limits. For example, it is perfectly fine to touch a friend on their head, arms, and upper torso. These areas are less appropriate to touch, however, if the person is an acquaintance or stranger. For them, only hand touching was deemed acceptable. Most of their bodies, especially for a male stranger, are considered completely off limits.

What can we do, then, to satisfy our needs for physical intimacy? Even in nonpandemic times, the areas through which we move on a daily basis—our workplaces, grocery stores, and coffee shops—are physical affection deserts. Even if physical affection is not specifically forbidden, social prescriptions discourage public displays of affection, especially affection directed toward strangers and coworkers. For many of us (in nonpandemic times), we get almost all of our physical touch needs met while in our homes. Sex, kisses, hugs, cuddles with humans and pets, and caresses—adults who work outside the home have a window of about twelve hours to get this type of physical affection on an average workday. Then once they walk out the door, this window closes. But what about those people who don't get much (or any) physical affection from their family members? And what about those who live alone? Where do they fulfill their needs for physical touch?

NUMBER ONE RULE: PAJAMAS STAY ON

Massages are one potential solution. If the world can be divided into camps of massage lovers and haters, I might be elected president of the massage-loving group. I'd certainly put forth a solid campaign. In nonquarantine times, I get them weekly, and I consider massages an essential part of my health maintenance. During quarantine, I felt the loss. I felt it in my hands, shoulders, and more generally, spirit. Today my massage therapist, Julie, mentioned that many of her clients are elderly. "They don't even come because they are in pain," Julie said, "they come because they are lonely." Weekly, biweekly, and monthly. They come to be touched in a positive way by another human. "One older man even told me that he has depression, and a massage is the only thing that works to make him feel better," she added. This man gets a massage twice a week.

As a psychologist, I must provide the caveat that scientific study endorses the benefits of psychotherapy for the treatment of depression.[25] That said, there is still much debate about the effectiveness of antidepressant drugs and growing concerns about their long-term use.[26] Perhaps because of this, people may be seeking alternative ways to cope with loneliness, anxiety, and depression. And although massage may not be a cure-all for all types of woes, there are a host of therapeutic values of touch. Some of these benefits relate to complementary therapies offered to patients with cancer, dementia, chronic pain, and those with other palliative and health care needs. Massage therapy is also sometimes pursued as an alternative treatment, as with Julie's depressed client. Is he the only one who has discovered this magic?

Decidedly, no. Around the world, doctors and scientists have been exploring the value of massage therapy for hundreds, if not thousands, of years. In terms of published scientific evidence, almost twenty years ago (in 2004) a meta-analysis of thirty-seven studies showed that even a single session of massage therapy had a significant positive effect on the body, including lower blood pressure, decreased pain, decreased anxiety, and lower heart rate. And over time, massage therapy reduced anxiety and depression, with treatment effects being similar in magnitude to a course of psychotherapy.[27] More recently, researchers have demonstrated that they can successfully remediate symptoms of generalized anxiety disorder with only six weeks of twice-weekly Swedish massages.[28] Researchers have not yet settled on an answer to why massage works. It might be that it promotes parasympathetic nervous system activity, improves circulation, or helps with restorative sleep. Or it might also be the attention from another person—specifically, their attentive touch—that provides the benefits.

Julie's elderly clients are not the only ones who have discovered the magic of massage. According to a Massage Therapy Association survey, this is a booming consumer market. More than 47.5 million people got massages in the United States in 2018, or about 21 percent of the population, and those who did had an average of 4.5 massages. A more interesting statistic is the rise in demand for therapists. From 2008 to 2018, massage therapists in the United States increased in number by 25 percent. And the demand is still rising drastically. The US Bureau of Labor Statistics, the government office that makes projections about shifts in the US labor force, predicts that the next decade should bring a further 29 percent increase in job opportunities for massage therapists.[29] For comparison, the projected increase in demand for all occupations over the next ten years averages only 4 percent, and the projected increase in demand for psychologists comes in even lower at 3 percent. With rates of mental illness and suicide plateauing, and even climbing, despite greater attention as well as access to psychological treatment, these trends suggest that people may be attempting to find new solutions for their woes.[30] Could the era of talk therapy be over? Perhaps we are entering a new era when touch reigns supreme.

And touch from professionals is not limited to massages. A new career field has emerged in the last couple of decades: professional cuddlers. For $149, people who have a panache for warm embraces can enroll in an online certification program to become a trained Cuddlist through cuddlist.com, which is currently offered to US residents only, but is gearing up for global expansion. After completing basic training, cuddlers can even become Certified Cuddlists through an evaluation and mentoring session with a Cuddlist expert and two reviews from clients. With going rates of $60 to $80 per hour for an average Cuddlist and rates up to $160 for a session with famed Cuddlist Yoni Alkan, the cost of cuddling is similar to that of a massage in the United States.[31] The goals are strikingly different, however. According to Cuddlist Training cofounder Madelon Guinazzo, Cuddlists are trained to give "healthy, consensual, touch" in the form of nonsexual cuddling.[32] There is no kneading, stretching, or manipulation. Instead, cuddlers offer their clients spooning, hugs, and hand-holding. It is an entirely different type of experience, but it also appears to be in demand.

Since 2004, the website cuddleparty.com has helped facilitate organized, nonsexual cuddling parties worldwide. As of 2021, the company has helped set up cuddle parties in the United States, Australia, Canada, Denmark, England, Sweden, and South Africa.[33] At these events, individuals cuddle in their pajamas with others, often strangers, in apartments, homes, yoga studios, or other cozy event spaces. Yet it is not a free-for-all cuddlefest. The number one rule is that pajamas stay on. Additionally, likely so that there is no unwanted sexual touch, there are strict rules around communication and consent. Attendees must ask permission to cuddle or touch another person, such as, "Can I spoon you?" Boundaries must be respected, or a facilitator will step in. Starting as low as $10 for entry to a cuddle party depending on the venue, the cost is not exorbitant, but the benefits of the physical touch can be great.

Physiologically, the flood of dopamine and oxytocin one gets from a hug exists whether the hug is from a stranger, friend, or loved one. Still, the amount of intimacy, desire, or love one feels with the person might determine the magnitude of that flood. Think back to a first hug or kiss with a romantic partner, and compare it to the hugs you get from friends, or the hugs and kisses you get from your partner one year, two years, and twenty years down the road. Desire matters. Intimacy matters. And *habituation*, the decrease in response that happens after continual exposure, can affect both. In other words, a hug feels good, but a new hug or a hug from someone you love and desire intensely feels even better.

Unfortunately, the pandemic habituated us to a lack of intimate touch from others. Fortunately, this means that we are now primed for hyperstimulation, and postpandemic touch will feel novel and invigorating. It's as if we all emerged from a long hibernation of isolation into the warm embrace of connection with others. Like Bambi.

THE GREAT PRINCE OF THE FOREST EMERGES

Bambi was the first movie I saw in a theater. One of my distinct memories is the stark contrast between the sad, dark winter when Bambi's mother is killed and the cheerful, colorful spring when his fellow woodland friends emerge amid fragrant patches of flowers. I can't help but draw parallels between *Bambi* and the pandemic, which though stretching more than a year, felt like one dark winter. Moreover, the timing of the vaccine administration meant that in spring 2021, some of us started to emerge from our pandemic shelters. Many of us were doing so tentatively at first. Because just like the traumatized munchkins in *The Wizard of Oz*, we had to first assure ourselves that the Wicked Witch of the East really was dead. But we did emerge. And the "spring" we emerged to felt like a lifeboat.

During the pandemic, we learned valuable lessons about basic human needs. We also learned about the value of technology for sustaining our most important relationships. As in other times of crisis, like the 2016 Pulse nightclub shooting in Orlando, Florida, where many of the club goers secured their phones before they did anything else, phones were rated just as important to survival as access to medicine for chronic illness, and almost as highly as food and medical care.[34] This trend aligns with assertions from communication theorists that phones can be companions during trauma as well as extensions of our social and emotional selves.[35] During this time, people also used social media at unprecedented rates to gather information and connect with others, thereby serving to both quell and heighten anxieties, strengthen some social ties and weaken others, and create new models for responsiveness and support.

In the end, the COVID-19 crisis unwittingly ushered in a new era of human functionality; our forced reliance on technology to navigate friendships, work commitments, health care, and even death stretched our imaginations regarding the role technology can play in interpersonal relationships and the global economy.[36] Moreover, it made us revisit whether restrictions on work and leisure travel might be a solution to the world's environmental crisis, echoing arguments by US-based philosopher Aaron James.[37] Rather than the drastic reduction of the workweek to twenty hours as proposed by James, though, perhaps technology-supplemented models of work, like we adopted during the pandemic, can help us avoid catastrophic environmental disaster.

Aside from the potential benefits of these changes in infrastructure, we can garner much from the perspective we've gained from this harrowing experience. Perhaps the isolation and division of the pandemic will be the ultimate learning experience, helping us to direct our focus and attention to the people in our world we cherish most. Certainly, many of us emerged from the pandemic stressed and anxiety ridden, but the solutions to this might be found in connecting with others. Free from the restrictions of social distancing, opportunities abound for touch, love, and affection. The pandemic, then, may prompt a rebound effect, ushering in a renewed era of intimacy and serving as a continual reminder for all the world's people to love one another unabashedly.

That was my advice to Sam, as she lamented over her losses, including the devastating loss of her father. "Wait for spring," I said, "and be prepared to breathe again."

HOW TO SURVIVE A PANDEMIC

And Now I Think about Nothing

Survival tip #1 Pandemic losses are not all tangible. Certainly, some people lost money. Some people lost jobs. Some lost people they loved. But almost all of us lost less tangible things, like time, opportunity, and hope. It's really easy to see the tangible losses, but the others are not as visible. Reflect on what you lost and encourage the people in your social spheres to do the same. Only then can we properly mourn our losses, heal, and move on.

Teachings of a German Psychologist

Survival tip #2 Depending on where you are in your life, the pandemic might have hit you in a different way. For some of us it was a soft tap, and for others it was a devastating blow. Don't compare your experience to anyone else's. Comparative grief is nonproductive. Some people prospered. Others suffered excruciating losses of people, jobs, and moments they cherished. Validate both the gains and setbacks. There is no way to rewind the clock. We all must trudge forward. But some of us will be taking postpandemic steps with heavier crosses to bear. Don't expect anyone to move at your pace.

Pandemic Puppies and the Upside of Voids

Survival tip #3 You control how you fill your voids. You can choose to do so with food and alcohol, or you can listen to your body about what you really need. Is it affection? Is it the warmth of another body? Company? If so, humans are not the only sources of connection. Pets can provide similar feelings of warmth and affection, and for many, are valued members of their family. If you're feeling lonely, consider petting or talking to a pet. If you don't own a pet, visit a shelter or take the neighbor's dog for a walk.

Has Anyone Hugged You Today?

Survival tip #4 Hug people. If someone needs social support, hug them. If someone doesn't need social support, hug them. If you are hesitant about their comfort level, ask simply, "Can I give you a hug?" Aim for one hug a day of at least twenty seconds, and work up to three per day as you gain more confidence and it becomes part of your way of life. Wrapping your arms around someone else's body in a (consensual) warm embrace might give you both more resilience and a healthier life.

Number One Rule: Pajamas Stay On

Survival tip #5 Acknowledge that touch is a basic human need. You are not weak if you need it nor indulgent if you seek it. Massages and cuddles are not just for those in pain or touch deprived. In fact, they might be the secret to better physical and mental health, even for those of us who are already healthy. Open yourself up to new touch experiences and try to surrender to the moments. In a world in which massage therapy is affordable and cuddle parties exist, there is no reason for people to be deprived of physical touch.

The Great Prince of the Forest Emerges

Survival tip #6 After the Persian Gulf War, a US special task force was appointed to examine how to deal with the stressors of war and challenges of reintegrating soldiers back into their families.[38] One of its most practical pieces of advice was to set small goals, disentangle the parts of the problem, and reward yourself for small victories. As you are still adjusting to the postpandemic world, take stock of your losses (physical, economic, and psychological), and aim for a slow reintegration back into your social and work life. Focus on one step at a time. Give yourself time to adjust. And reward yourself handsomely for each step. You have already survived; now is the time to thrive.

Bonus Technology Tip

Technology was a connector, refuge, and even lifeline for most of us during the pandemic. We learned new ways of learning and interacting, and developed routines around technology use that were necessary and beneficial. Keep those and use them to grow. Postpandemic, however, we must also make attempts reengage with our physical world in intimate spaces with meaningful others. This shift may feel unfamiliar and even scary. Give yourself grace as you move forward.

2 HOW TO SURVIVE CHILDHOOD

DON'T HUG OR KISS YOUR CHILDREN?

As an undergraduate at Cornell University in 1994, I worked in a developmental psychology lab for Elliott Blass, studying how to soothe infants without touching them. The route chosen in our investigations was a dilute sucrose solution, delivered via a needleless syringe or pacifier to babies from only a few hours to twelve weeks of age. The sucrose solution worked remarkably well; during painful procedures like circumcision or the heel prick to test for phenylketonuria, an inherited metabolic disease, babies who had been given sucrose, as compared to water or nothing, cried less.[1] The research was exciting, and I felt privileged to be part of a laboratory doing groundbreaking work. But my most memorable experience from those years was entering the hospital rooms of new mothers, asking them if they wouldn't mind if we gave sugar through a syringe to their new, perfect, have-never-eaten-anything babies.

There were so many cringeworthy aspects of this appeal. Sugar. Syringe. A perky young stranger in a white lab coat bopping into a hospital room asking a mother who just gave birth to do an experiment on an hours-old baby. I am still shocked to this day that so many mothers said yes. I did have the esteem of Cornell behind my appeal, and I'm sure that helped. Still, what helped more than anything was the expected benefits from the study. "This study may not benefit your child," I'd say, "yet the knowledge we gain from this project may help us improve care for infants who undergo painful procedures but cannot be touched, such as those in neonatal intensive care units." It was a nearly irresistible plea. Mothers, having just held and cuddled their newborns, knew the value of touch for soothing, growth, and bonding, and they were willing to allow their own newborns to participate in our sugar analgesia experiment on the chance that it could help a baby deprived of touch.

It's been nearly thirty years since I took part in this research, and much has changed in the meantime. Providing sugar solutions for babies undergoing painful procedures has fallen out of favor. Likely due to increased research and exposé films like *Fed Up*, the entire sugar industry is facing increased scrutiny and criticism. Instead, skin-to-skin contact, massage, and other types of positive touch for infants are strongly endorsed. Doctors are prescribing touch as a treatment for children born with a wide range of issues, from prematurity to neonatal abstinence syndrome (NAS; indicating withdrawal from drugs exposed to in the womb).[2] And with the uptick in NAS over the last couple of decades, due largely to the much-advertised opioid epidemic, finding effective solutions to soothe infants in withdrawal has become an increasingly pressing concern.[3] In the United States, for example, a baby with NAS symptoms is born every fifteen minutes, representing a 433 percent increase in the incidence rate from 2004 to 2014 alone.[4] From vibrotactile stimulation to volunteer cuddlers, doctors have been trying innovative solutions to calm these inconsolable, fist-clenching, NAS babies, who often have trouble sleeping.[5] The value of skin-to-skin contact, especially among infants, has been proven consistently valuable across a range of studies.[6] Touch, we've found, is healing, even for those difficult to heal.

Even more has changed in the last hundred years regarding expressions of affection toward children. In 1928, psychology's own John Watson declared in his book titled *Psychological Care of Infant and Child*, "Do not hug or kiss" your children. "Never let them sit on your lap. If you must, kiss them once on the forehead when they say good night. Shake hands with them in the morning. Give them a pat on the head if they have made an extraordinary job of a difficult task." Today, we know that affection from and toward infants and children can help solidify attachment with caregivers, laying the foundation for children's ability to express affection to others as they grow older.[7] Hugging and kissing your children is no longer eschewed; instead, it is encouraged as part of a normal, healthy developmental process. And yet when I ask my students in my introductory psychology classes to raise their hands if they remember *ever* being kissed or hugged by a parent, only about half of them do so.

This disconnect between what we should be getting in order to develop into intimate beings and what we are actually given in our home environments exemplifies a point that will be repeated throughout this chapter: surviving childhood is not really on the shoulders of children. It's really all about parenting.

WOULD YOU RATHER HUG A CLOTH OR SUCK MILK FROM A WIRE?

In March 2010, Australian mother Katie Ogg had just given birth to premature twins, born at only twenty-seven weeks.[8] One of the twins, Emily, survived. But after the medical team struggled trying to revive the boy twin, Jamie, for more than twenty minutes, the doctor pronounced him dead. The nurses left the deceased child with his parents, laying him across Katie's bare chest, so they could say goodbye. Devastated, Katie and her husband, David, began talking to their son, rubbing him, and telling him goodbye. After about five minutes, Jamie started moving in startled, short movements. Just reflexes, the doctor assured them; their son was not alive. After about two hours, Jamie opened his eyes. He suckled some breast milk Katie put on her finger, and wrapped his tiny hand around Katie's and David's fingers. Thinking their baby was showing signs of life, they urged hospital personnel to summon the doctor. The staff dismissed their urgings, however, insisting that the baby had not been revived back to life. The Ogg's tried a last plea, indicating that they had accepted their son's death and just wanted the doctor to answer some final questions. When the doctor returned, he took out his stethoscope and found a heartbeat. The child was alive. As of the latest reports, Jamie is a thriving, healthy little boy.[9]

Although an extreme example of the value of skin-to-skin contact, the story provides a tangible illustration of one of my core beliefs: *touch is invaluable to life*. But it is not only me who believes this. I have some heavy-hitting, renowned psychologists in my camp.

In the 1940s, Austrian-born psychiatrist René Spitz reported on his groundbreaking work with institutionalized infants.[10] The goal of his work was simple: measure the developmental trajectories of babies in two different institutions in South America. Institution #1 was a prison in which the infants were kept in a nursery and raised primarily by their incarcerated mothers, who had daily contact with them. Institution #2 was a "foundling home" where babies were raised by overworked orphanage personnel, each of whom had eight to twelve infants in their care. In both institutions, infants received adequate medical care, food, and hygiene. The sole difference between these two institutions was the amount of emotional support the infants received from their primary caregiver. Over the five years during which Spitz and colleagues observed these two groups of children, there were remarkable differences, which started to emerge early, when the infants were four to five months old. Those in the nursery group were smarter and more advanced in key areas of child development, like memory, mastery of bodily functions, and social relationships. As they grew older, the two-year-old children from the nursery group developed into normal functioning toddlers, while many of those in the orphanage had not yet learned to walk or speak—two key developmental milestones of this age group. Yet the most compelling results from this study relate to infant mortality. In the first two years of the study, not a single child in the nursery group died, but in the foundling home, 37 percent of the infants had died, mostly in the first year of life. The children in the orphanages also exhibited extreme levels of distress or in some cases apathy, suffering from what Spitz termed *marasmus*, a condition marked by physical and emotional decline in those starved of emotional stimulation.

Following up on this work, psychologist Harry Harlow conducted groundbreaking social isolation experiments with infant rhesus monkeys in the late 1950s.[11] Harlow separated newborn monkeys (just a few hours old) from their mothers and raised them in isolation for anywhere from three months to a year. Much to the chagrin of modern-day psychologists, who are infinitely more attuned to animal welfare in experimentation, the monkeys raised in isolation suffered. When integrated with peers, the once-isolated monkeys exhibited minor to severe disruptions in their social development, such as apathy, awkwardness, and fear. Harlow also tested conditions in which the baby monkeys had the option to cling to cloth or wire monkeys, the former of which offered contact comfort while the latter offered food. When given the choice of a soft monkey to cuddle or a mesh wire monkey delivering food, the infant monkeys clung to their surrogate cloth mothers for hours and hours each day, opting for contact comfort over food. This led Harlow to make one of the most notable discoveries in developmental psychology: contact comfort is crucial for healthy development. My mom agrees. Vehemently.

WHAT MY MOM AND TOM BRADY HAVE IN COMMON

Over the past few decades, I've had thousands of conversations with people about the way they were raised. Having reflected on these unique data points, I can now say definitively that my upbringing was nontraditional. There are two stories I like to tell that illustrate this well. The first relates to my childhood home. I moved more than fifteen times before I was twelve. My parents, self-pronounced hippies, drove west in the early 1970s, looking for a place to live that felt good. They eventually landed in a tent by a river in Colorado, enjoying the benefits of vast fields of wild grasses and fresh, open air. I don't have memories of this time, but I can imagine my mom leading us by the hand to bathe in the river, making daisy-chain wreaths for our hair, and rocking us to sleep in her arms as she sang Judy Collins songs. My mom felt good there. Until she suddenly realized that my little sister and I might be eaten by bears, and demanded we get a real house immediately.

But that house didn't last long, and my mom, a rolling stone that definitely gathered no moss, moved us from place to place until we finally settled when I was in middle school. My early childhood could definitely be described as chaotic. Which leads to my second story. Like millions of children worldwide, I had the stability of *Sesame Street* to keep me grounded. This in itself is not unique. Premiering in 1969, *Sesame Street* has become one of the most popular television shows of all time. The unique part of the story is that while I was watching *Sesame Street* in the morning, my mom would give me some sort of bizarre, healthy breakfast. I sometimes have flashbacks to these moments. Five-year-old me, sitting cross-legged, watching Grover go near and far, eyes wide with interest, holding some atypical breakfast treat. A rice cake. A tomato. Or half of an avocado.

It should have been no surprise to me, then, that when I had my own children, my mom insisted on two things: homemade baby food and infant massage. Although I will leave the baby food recommendations to the nutritionists, I will say it was surprisingly easy to boil and blend fruits and vegetables, and spoon them into ice cube trays. This book is not about a food famine, however. It's about intimacy, and as my mom suggested, massaging my sons had a profound effect on them. And me.

I remember being unbelievably exhausted when, as the mother of a days-old baby, I first watched a how-to video on infant massage. I made an effort to remember the key components. First, ask the baby, "Would you like a massage?" The baby never responded, but I guess this was an early lesson on the importance of affirmative consent. Second, massage with only a couple of fingers in a clockwise, circular motion over their belly, and spell out ILU. I love you. Third, make eye contact and make it a pleasurable experience. Of all the purported benefits of infant massage, the one I clung to most through the hours I spent massaging my two infant sons was that I was communicating my love of them through touch.[12] In this time of rapid brain development, I was making sure their hardwired connections included soothing words from a loving mother, a calm, warm environment, and soft caresses. I also became more in tune with their needs; I could tell when they relaxed. I learned when they wanted more and when they wanted the massage to end. I'm convinced that these massages made them better babies and made me a better parent. Most important, they helped my two sons appreciate the benefits of physical touch, remnants of which I still see today.

At a swim team banquet last year, one of my son's teammates approached my then eleven-year-old. "Did you see Joseph?" he mocked. "He was holding his mom's hand!" My son looked at him, somewhat bewildered. "So?" he asked and walked away. This young boy's sentiment reflects a discontinuous cultural message. Physical affection between children (especially boys) and their parents is often mocked or discouraged, but as these children grow into young men and women, they are supposed to be affection aficionados. This was salient to me in 2018, when US football quarterback Tom Brady was harshly criticized for kissing his eleven-year-old son on the mouth. The internet broke into outrage. A father kissing his son on the mouth challenged what we know and expect about parent-child affection. Or did it?

Parenting style can be broadly characterized along two spectrums: warmth and control. Authoritative parents, those who balance a high degree of warmth with a high degree of control, have children with the best outcomes. Their children are more secure and successful, based likely on the fact that they had structure and rules, but also love and support. Parental affection, such as hugs and kisses, is firmly in the camp of warmth. It's a demonstration of care, love, and attention. And provided the child does not perceive the affection to be excessive, expressions of affection are beneficial to healthy development and positive relationships.[13]

You see, as children we develop the *internal working models* of attachment that we use as reference points for all of our future relationships.[14] The relationships we have with caregivers matter. They teach us about whether to trust others. They give us esteem. They help us understand our value in the world. Yet the working models we develop about attachment relate not only to our security in our relationships. We also develop models of interaction patterns. We learn, through imitation, how to love others, and how to express affection and intimacy. By kissing his son on the mouth, Brady was expressing love to his child and promoting a framework for emotional expression that his son could use to navigate his future relationships. As a child who still kisses her mom on the mouth, I want to say this to Brady: thank you. Not for leading your team to six Superbowl championships, but for holding strong against a world of critics, and reminding everyone that we no longer live in a world in which we are discouraged from hugging and kissing our children. My mom, forever a hippie, making daisy-chain hair wreaths and handing me avocados, thanks you too.

MISTER ROGERS VERSUS *SPONGEBOB*: WHO WOULD WIN?

Reflecting on my early experiences really helps to put the span of the last forty years in perspective. My mom stayed home with me and my three siblings—a trend that's much less common now. Almost twice as many women are in the US workforce today than in the 1970s and early 1980s, and four times as many women are in the workforce now than in 1950.[15] This has meant a massive shift in the ways in which families function. With both parents working, the strain of home life is felt by both mothers and fathers, more than half of which state that working makes it more difficult to be good parents.[16]

Perhaps to meet this need, the entertainment industry has seen huge expansions in media offerings for children. When parents get home from a long day at work, still needing to prepare dinner or maybe finish some last bits of work, television is a popular babysitter. Long gone are the days of single offerings of *Sesame Street* and *Mister Rogers* through public broadcasting stations. Families no longer rush home to watch the once-a-year showing of *The Wizard of Oz*. Instead, in many countries, children have access to hundreds of children's programs, livestreaming throughout the day, or available on demand via open sources like YouTube or on pay networks like Disney+. In the talks I've given across the country, I often weigh what we've lost in entertainment versus what we've gained.

I recently gave a talk to more than five hundred Head Start teachers in Indiana. Head Start is child enrichment program founded in 1965 by US president Lyndon Johnson to help give low-income preschoolers the educational and nutritional foundations they need to have a good start in their first year of school. While speaking to these teachers, I engaged in a lively discussion about the value of media for shaping young minds. We centered much of our conversation on the television show *Mister Rogers' Neighborhood*. Mister Rogers, in his cardigan and house sneakers, showed us that we could express our feelings, and that it was OK to be sad, scared, or shy. He talked about difficult topics, like divorce and war, and took us with him on outings to show us how things like crayons were made. He was also a champion of racial equality and inclusion. Many people remember him inviting Officer Clemens, a Black police officer, to join him by putting his feet in a pool, breaking the segregation barriers that still existed in the late 1960s.[17] One of the teachers also mentioned his memorable broadcast in 1981, when he had a young boy in an electric wheelchair explain his medical condition and how his wheelchair worked.[18] It was the first time she had ever seen a child with a disability, and she was forever moved. Despite all of these valuable lessons, and the show's indelible impact on the social and emotional development of millions of children, *Mister Rogers* stopped airing on PBS in 2001.

In that same year, Disney acquired the Baby Einstein franchise from entrepreneur Julie Aigner-Clark for an undisclosed amount. Marketed as educational resources that would make children smarter, Baby Einstein videos were snatched up by well-meaning parents, apparently unaware of the American Academy of Pediatrics' recommendation that children under the age of two should have zero screen time. The franchise quickly crumbled, however, as researchers began to demonstrate that children did not learn anything from these videos, and experts doubled down on their recommendations that children under two should not watch media of any sort. A claim was filed with the Federal Trade Commission, and Disney was forced to change the labeling of the videos, taking off claims that they were "educational."

The collapse of the multimillion-dollar Baby Einstein franchise was pivotal in our culture. Parents worldwide had been convinced that these entertaining and colorful videos were going to make their children geniuses, and when science showed that their educational claims were unfounded, Disney was forced to issue refunds globally.[19] This is a stellar example of how advertisements can deceive the masses, but from the perspective of technological innovation, the Baby Einstein debacle also exemplifies how fast-paced technology adoption, without scientific guidance or appropriate public policy, can be detrimental to human development. Psychologists have long known that babies do not learn well from passive video consumption. And although we now understand that live, interactive video might be somewhat beneficial, a human companion (even in addition to live video) is the best source for learning new material, especially language.[20]

Considered together, these events were the starting waves of a cascade of changes in children's media consumption around the world. Mainstays in the educational television programming market, like *Sesame Street*, also began to lose popularity as children shifted their focus to shows like *SpongeBob SquarePants*. When I asked the five hundred Head Start teachers if any of the preschoolers in their classes could name all ten of the popular *Sesame Street* characters I put up on the screen, no one raised their hand. Not a single one. Incredulously, I asked them what characters these children knew. "*Fortnite* characters or YouTube stars," they said. *Fortnite* characters were now garnering more attention than Grover. This was a big stab to my heart as I remembered how energetically Grover ran back and forth across the screen, teaching me near and far.

THE PIED PIPER OF SILICON VALLEY

The entertainment industry has not only expanded its offering for children. Parents too are spoiled for choice in their media consumption. Television, Netflix, the internet, email, text messages, and social media—they are irresistible sirens in a sea of mundanity. Lured into their online worlds, parents are also captivated, inadvertently modeling distancing behaviors in the course of their everyday lives. Visit a restaurant or park, and look to see how many parents are on their phones. It is not uncommon for me to go out to eat and see all of the members of a family on their devices. Technology addiction among parents is such a significant concern that Common Sense Media, a nonprofit organization that promotes and advises on the safe use of technology and media, released a marketing campaign focused on device-free dinners. In one popular video, actor Will Ferrell plays a father engrossed in his phone while his children make shocking admissions to get his attention. His teenage son jokes that he is selling bongs out of their minivan, and his young daughter claims to be cooking meth in the basement. All the while, Ferrell's character is barely paying attention, acknowledging their admissions with inappropriate approval. The comical skit confronts the issue with humor, but the problem is serious. Parents addicted to their devices are raising children addicted to their devices.

Cognitive developmental theorist Jean Piaget is widely cited as the founder of constructivism, a theory that posits that we construct knowledge from experiences and that these experiences shape our cognition. Psychologist Lev Vygotsky expanded on this model, suggesting that this process is inherently social and our thinking is shaped by our experiences with others. Both theorists agree that imitation is an important milestone in cognitive development; when children are able to mimic those they see as competent (e.g., their parents), they derive self-esteem from those actions. Then through a series of punishers, reinforcers, and vicarious learning experiences, children learn how to function in the world as social beings.

As parents navigate social communication through technology, two simultaneous processes have the potential to affect development. First, research has shown that there is the potential of technology interference across a wide range of parenting behaviors, especially when mothers are depressed and addicted to their phones.[21] This can have a significant effect on parental responsiveness, which has been shown to be essential to healthy attachment to caregivers and a host of developmental outcomes.[22] Second, as children watch their parents interact with technology, they learn how to navigate their social worlds in the same way as their competent caregivers. They are spending more time online, and this, according to leading world experts, may lead to significant declines in physical as well as emotional health and wellness. This has become such a crucial issue for child development that groups concerned with the welfare of children, like the American Academy of Pediatrics and World Health Organization, now advocate for limited media use, more exercise, and more outdoor play.

Even as I type these words, I'm in disbelief. I am shocked that I live at a time in history when leading experts in health and child development have to advocate for outdoor play. These guidelines do, however, address a trend I've been following closely for the past few years. I casually refer to it as the "no kids at the pool" phenomenon. We live in a subdivision purpose built for families. With more than fifteen miles of walking paths behind our houses, a golf course, tennis and basketball courts, parks, and a collection of pools, my neighborhood was deemed by local realtors a perennial favorite for young families. Nevertheless, excluding the momentary uptick in traffic I've seen during the pandemic, the paths and parks are largely empty, and the pools, once packed with kids playing sharks and minnows, are deserted. A few years ago, I heard a young teenage boy strike up a conversation with a similar-aged girl. "Where's your sister?" he asked. "Oh, she always stays home now," she replied. "She's on her phone all day." Eventually the teenage boy stopped coming too, as did the sister.

I grew up in an era where outdoor play was the sole summer pastime. Pool excursions, showcased in films like *The Sandlot*, were highlights of a long, lazy summer spent playing outside with friends. As a mother of two preteen boys, I can attest to the fact that life has changed drastically. Even in my family-oriented neighborhood in a small Indiana town, the streets are fairly empty. It's as if the Pied Piper of Hamelin swept through, playing his pipe and luring the children into the caves of their homes. Except, in this case, the Pied Piper is no stranger. We know him. He lives in Silicon Valley, and almost every single one of us has been lured into a cave by his tune.

Almost every one of us, except the Pied Pipers themselves, that is. As highlighted by popular news outlets, some of the biggest leaders in technological innovation, the same Silicon Valley executives creating the media that leads us into our caves, eschew media for both themselves and their children.[23] Founder of Square and Twitter CEO Jack Dorsey, who is known for walking to work and taking meditation retreats, famously uses the screen time function on his iPhone to limit his time on Twitter to only two hours per day.[24] And the tech giants' rules around children's use of media are often much stricter. The most recent World Health Organization guidelines recommend less than one hour per day of media use for young children under five, yet many of these technology leaders restrict their own children's use to far less than that.[25] For example, it has been reported that Evan Spiegel, the creator of Shapchat, limits screen time for the seven-year-old son he shares with model Miranda Kerr to only ninety minutes per week.[26] Microsoft Founder Bill Gates and Apple cofounder Steve Jobs famously placed limitations on their children's technology use, with Jobs apparently not allowing his children to have a phone until they were fourteen.[27] And the Waldorf School of the Peninsula, a Los Altos, California, private school favored by some of Silicon Valley's tech moguls, has a media and technology philosophy that eschews the use of media for young children. According to its website, electronic media introduced to children and adolescents "can detract from their capacity to create a meaningful connection with others and the world around them."[28] The Pied Pipers, it seems, have wrapped blindfolds around their own children's eyes and stuffed cotton in their ears, no doubt to ensure their children won't be easily led by the seductive tune.

WAYFINDING ACROSS THE PRAIRIE

News reports highlighting stories like these often incite furor. Despite the fact that the American Academy of Pediatrics, World Health Organization, and other child advocacy groups have cautioned against extended screen time, the technology industry has continued to produce hardware and software aimed at children. According to market research, households with children drive technology market consumption; they are the innovators and early adopters of internet services and smart-home technology.[29] Advertisers and marketers of technology also target children, specifically, as a powerful source of consumerism in modern culture.[30] How dare these technological giants create tools for mass consumption that are potentially harmful for children? In talks I give in schools, parents frequently approach me with this question. They feel cheated and deceived. In response, I offer a gentle reminder of basic economics. "If there were no demand," I assure them, "they would no longer be supplying them."

These tech executives should not be admonished for their actions; they should be modeled. They have found innovative ways to address basic human needs for entertainment, socialization, and content creation. This is no crime. And if we were to criticize them for these innovations, shouldn't we also criticize the likes of Benjamin Franklin, Henry Ford, and Alexander Graham Bell? As with many things, the inventions themselves are not the problem. It is the way in which society uses the invention that has created the problem. But the problem doesn't affect us all equally. These tech executives are often constructing lives for themselves and their families that capitalize on the benefits of media while avoiding the pitfalls. Their key advantage over the average person is the way they are using their knowledge. Likely privy to the usage trends in tech along with the research related to its benefits and detriments, they use this knowledge to carefully curate an enriching life—one not dominated by technology addiction. Yet it is not only the tech executives who know about the potentially damaging effects of excessive technology use among children.

We have reached a turning point in our history regarding children and screens. We now know, definitively, that extensive technology use among children can be detrimental. The American Academy of Pediatrics has told us. Common Sense Media has told us. The World Health Organization has told us. There are no mixed messages here. We know. Reflecting this, a 2020 Pew Survey of US adults showed that 71 percent of parents with a child under the age of twelve indicated that they were at least somewhat concerned that their child was engaging in too much screen time, and 31 percent indicated that they were very concerned.[31] And although the vast array of technological innovations might appear on the surface to make life easier for parents, the same 2020 Pew Survey showed that two-thirds of parents say that parenting is more difficult now than it was twenty years ago. The culprit according to parents is technology.

To be fair, this knowledge spread unequally throughout the population. Silicon Valley executives, who likely saw the negative effects of technology use early and firsthand, may have been the first wave of those eschewing screens. As evidenced by the record-breaking attendance at my digital wellness talks in private schools, educated and wealthy parents quickly followed suit. These parents are mobilized for a backlash—a rebound back to simpler, nontechnological times. They want their children to learn knitting, baking, and wayfinding across a prairie. "Tell me that I need to take away all technology," one of these parents told me recently, "and I'll do it." Lower-income families, whose kids are just now getting one-to-one technology in their public schools, appear to be the last to follow suit. In comparison to their higher-income peers, lower-income teens spend an additional two hours per day engaging with media (eight hours, seven minutes for lower-income teens versus five hours, forty-two minutes for higher-income teens).[32] The message that children must be technologically savvy in order to thrive in a modern world has been aimed at all income levels. Yet those in the lower-income brackets, perhaps afraid that their children might fall behind, seem still to be clinging to this message that their children must be constantly wired in order to be ready to tackle the technological demands of a new era.[33] This simply isn't true.

Echoing the sentiments of Dorsey, who famously noted that what someone learns online is more important than how much time they spend there, how much time children spend online may not be as important as what they do online.[34] Certainly, the enormous amount of time children spend online is concerning. In a 2015 Common Sense Media study, researchers found that excluding the hours using media for school or homework, US preteens spent about six hours using media every day, and teens spent about nine hours using media.[35] Media consumption has reached epic levels among teenagers, prompting experts to warn that screen time is now competing with sleep. These statistics are staggering. Even more concerning, though, is what they are doing online. Among both preteens and teens, more than one-third of this time spent online was passive consumption, like listening to music and watching TV and videos. Interactive consumption, such as playing games and browsing online, was next most popular, followed by communication via social media and video chats (which accounted for 14 and 26 percent of tweens' and teens' time online, respectively). Coming in last, at 3 percent of the total time spent on media for both age groups, was creation, which included writing, making digital art or music, and programming. In other words, teens are not spending time online learning to make pie charts or how to code. They are watching TV, kicking it with friends, and gaming.

As a child of the 1980s who remembers spending time on the phone talking with friends, playing *Pac-Man*, and watching *Diff'rent Strokes*, it's difficult for me to criticize a pattern that seems somewhat reflective of my own upbringing. Many of you might be thinking the same thing. Yet this is where our knowledge of proportions comes in handy. Think about how much screen time you had as a child. The balance of my life as a kid was not weighted toward computer-mediated interactions and passive consumption. The leisure I squeezed in between 3:00 p.m. (the time I got home from school) and 8:00 p.m. (my bedtime until I was a teenager) was likely divided somewhat like the pie charts I created below. In early childhood, I had little interaction with technology, likely due to the fact that there were no cell phones or internet, and we had three television channels on the single TV in our living room. Later in my teenage years, we had cable television, but my interests switched from life inside my house to spending time with peers outside my home. I cannot remember any screen time in my teenage years, aside from an occasional movie. And I even worked in a movie theater. On balance, my childhood was very much wayfinding across the prairie.

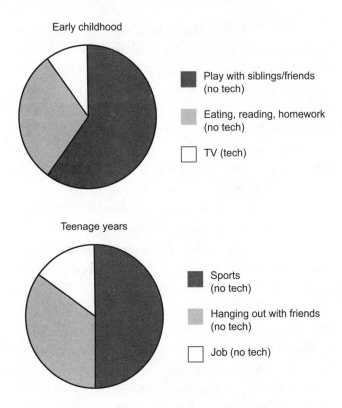

Early childhood

- Play with siblings/friends (no tech)
- Eating, reading, homework (no tech)
- TV (tech)

Teenage years

- Sports (no tech)
- Hanging out with friends (no tech)
- Job (no tech)

That said, social life on the prairie is different now. Instead of talking on the phone, kids text and talk on Discord while playing interactive video games. They DM each other on Instagram and send silly messages on Snapchat. And when they meet, they snap selfies and make TikTok videos together. All of this is social activity—different, of course, than the social activities some of us of the older generations engaged in when we were young. But they are social nonetheless, learning the methods of wayfinding necessary to navigate a tech-dominated culture. This includes being able to navigate a ruthless and unforgiving cancel culture, where online shaming and cyberbullying dominate the landscape. They must learn to step carefully through the land mines of trolls and sadists to find pockets of support, kindness, and avenues to thrive. This is the technological world we live in.

OBI-WAN: YOU'RE OUR ONLY HOPE

One of the most ambitious and in-depth longitudinal studies of human development is the Harvard Study of Adult Development, which was started in 1937 by Harvard professor and psychiatrist George Vaillant, and funded by US department store magnate and philanthropist W. T. Grant.[36] Now under the direction of Robert Waldinger, the study has been following a group of men recruited from Harvard and neighborhoods of Boston for more than eighty years. You might be familiar with the Harvard student part of the study (the Grant Study), which recruited 268 men from the Harvard classes of 1939–1944, when they were nineteen years old on average. The cohort had some notable participants, including President John F. Kennedy and Ben Bradlee, longtime editor of the *Washington Post*. The Boston cohort, in contrast, was strikingly different in terms of demographics. Also referred to as the Glueck Study, in reference to Harvard Law School professors Sheldon and Eleanor Glueck, who spearheaded the study, this investigation included almost twice as many participants (456) recruited between the ages of eleven and sixteen from different socioeconomically disadvantaged neighborhoods of Boston.

Despite differences in sample composition, the methodology of these two studies has been consistent and incredibly comprehensive. Every two years, participants completed exhaustive surveys, gave biological specimens like blood samples, and received brain and MRI tissue and organ scans once the technology was available. The study also expanded to include the offspring of men from the two original samples as part of its Second Generation initiative. It has been an enormous study of great significance, and its findings are notable. Of all the metrics the researchers have collected, and the pinpricks, prods, surveys, and scans, there is one measure that most of all seems to predict great outcomes: human connection and love. Friendship. Intimacy. Those who thrive differ from those who don't in terms of how close they are to others. And what effects has technology had on these participants along with their ability to develop and sustain intimate relationships? It's not yet known. TBD.

Reflecting on shifts in media consumption and how they might affect social connectedness feels especially relevant now. Like many parents worldwide, I found 2020 to be exhausting. Trapped in the house or yard with no social stimulation from peers appears to be quite a challenge for all children everywhere. For a large portion of the year, I spent most of my children's seven-hour school day watching them wander around the house, grab snacks, and hit golf balls in the living room. Thrust instantaneously into an e-learning context, my children's public school teachers appeared to be grossly underprepared to manage children at a distance. From a pedagogical standpoint, I empathize. I have been teaching online for fifteen years, and still there is nothing I have produced that is as engaging and informative as the two-minute algebra video I just watched on Khan Academy. Nevertheless, the juxtaposition of the subpar learning environment in which my children spent their days and the slick, commercially created online learning content (both paid and free) has me pondering the origins, present state, and future of digital education.

I can see the future of online education so clearly. It includes interactive virtual assistants leading children through a personalized curriculum. Scaffolded and developmentally appropriate assistance provided by a 3-D avatar, like Princess Leia when she makes her appeal to Obi-Wan to help fight against the empire in *Star Wars Episode IV*. A kind and ever-patient mentor, encouraging children through learning activities tailored to the child's interests. Live video connections made instantly with other children around the world, allowing cross-cultural collaboration as well as glimpses into the homes, schools, and lives of people in other cultures. Personalized online education is a promising solution to the one-size-fits-all approach that teachers have to adopt to meet the needs of overcrowded classrooms. The potential is boundless.

In contrast, the e-learning environment I opted them into had mostly static online assignments interspersed with occasional Zoom meetings and chats with classmates. During the past year, their school lives have been mostly social deserts. "How is this affecting them," I recently asked my husband, "having only each other and their parents as their sole social contacts?" He shook his head. He didn't know.

This thought prompted a shift in my structure around technology use. Usually quite stringent in my rules about screen time and building social connections on the internet, I started allowing my children to chat with friends while they were gaming. They texted their friends and watched YouTube videos together. And I realized something important: a social pool can be built in little drips. There is a popular theory in media that suggests that influences can be major, drenching us with information that makes us develop or change our attitudes. Alternatively, they can be minor, dripping into our worlds, subtly shaping our attitudes toward things like race and gender.[37] Perhaps today's kids do not need be drenched in two-hour phone calls with friends to develop the skills they need for social connectedness. Maybe they need only the drips of gaming conversations and social media comments to make and keep friends. Eventually these drips will evolve; emails to colleagues, text messages to friends, and video chats with lovers will become commonplace, overshadowing the digital footprints they left as teens.

This realization was pivotal, helping me put my own life in perspective. I'm no longer wandering the prairie with a compass and canteen. I have an iPhone and Hydro Flask that I ordered on Amazon. I make pie charts in Excel and am writing this book on a laptop. And I text my friends; I rarely call them. Our children do need to know how to use technology in today's world, and using it well might help them prosper in both their careers and social lives. Although they might be missing touch (from friends, at least), the drips of social connectedness they get from their life online, just like the drips of sucrose we gave to babies in the hospital, might just be enough.

HOW TO SURVIVE CHILDHOOD

Don't Hug or Kiss Your Children?

Survival tip #7   Your parents and grandparents were raised in a different era, where hygiene and nutrition were valued, and touching children was discouraged. The advice they give you about raising children may not align with what experts know today. What we now know is that touch is invaluable. Skin-to-skin contact, hugs, and massages—from the time they enter this world, affectionate touch has a profound effect on children's health and development. When a baby cries, pick them up. Hold them lovingly for as long as you can. The last time you ever hold your child will pass by as an ordinary moment. Only when you look back will you realize that period of your life is over.

Would You Rather Hug a Cloth or Suck Milk from a Wire?

Survival tip #8   Although neither a cloth nor wire frame sounds like a good companion to me, baby monkeys isolated from their mothers spent up to twenty hours a day clinging to a cylindrical cloth. Rocking themselves. Soothing their lonely hearts. Contact comfort appears to be paramount in the development of healthy relationships. Not only does it help with socialization, but intimate interactions help make children smarter. Give it generously.

What My Mom and Tom Brady Have in Common

Survival tip #9   You are the greatest source of love and intimacy your children may ever know. The affection you show them while they're young will have a lasting impact on all of their future relationships. If you want your kids to know how to love others, love them. And you don't need baby massages or kisses to communicate affection. We have all been influenced by the models our parents set for us, and not all of us are comfortable with close intimate contact. But you must give something. Find what works for you and do it consistently.

Mister Rogers versus *SpongeBob*: Who Would Win?

Survival tip #10   Mister Rogers may be my favorite character ever on television, and SpongeBob is probably my least favorite. Yet the hard truth is that Fred Rogers died in 2003, and *SpongeBob* is still airing on Nickelodeon. As Mister Rogers said, "You rarely have time for everything you want in this life, so you need to make choices. And hopefully your choices can come from a deep sense of who you are." Decide who you are and what you want your children to be, and make intentional choices about the media your children consume. Make every effort to make sure that you fill their young brains with shows promoting empathy, kindness, learning, and diversity. Donate to public broadcasting. Demand more educational programs for children. And if you want your children to stop watching YouTube stars, remember that you set the rules.

The Pied Piper of Silicon Valley

Survival tip #11   The biggest lessons you can teach your kids regarding their own technology use echoes sentiments from experts worldwide: don't let screen time interfere with exercise, outdoor time, or sleep. And if you want your children to really thrive, steer them away from selfies and toward Excel chart building. If the Silicon Valley executives can enforce strict rules around tech use, so can you.

Wayfinding across the Prairie

Survival tip #12 You cannot compare your children's lives to yours. It's just not fair. If you had three channels, one TV, and a set of encyclopedias, you have no idea what it's like to grow up in an era of TikTok videos and Netflix. Reflect fondly on the days of yore, but acknowledge the irresistible appeal of our tech-laden world. Using technology is easy and addictive. However, we now know that excessive technology use may be detrimental to children. If you found out your twelve-year-old was addicted to drugs, you wouldn't keep giving them the drugs. You'd take them away. Never be afraid to do the same with their iPhone.

Obi-Wan: You're Our Only Hope

Survival tip #13 During a period of quick development, you must expect some growing pains. We are currently in the awkward teenage years of technological innovation, full of angst and uncertainty. But hope is on the horizon. The best, most beneficial innovations—those that will help our children learn and socialize better—are still to come. It's an exhilarating time to be alive. For now, though, our children are OK. Even during a period of extreme isolation, they survived. Reshape your views on social connection. Accept that kids no longer need a compass to traverse the prairie. Provided you gave them warmth and structure, for your kids, Google Maps will work just fine.

Bonus Technology Tip

As you try to shape your child's world, technology will be a necessary consideration. How much? When? What type? Where? Before you place any device in your child's hands, make sure you have established rules around its use. And make sure you and partner, if you have one, are on the same page. Fights about technology between children and parents are intensified when parents aren't on the same page. Common Sense Media and the American Academy of Pediatrics provide some structure and information, but there are few hard rules. Improvise. Negotiate. But whatever you do, create some structure around device use.

3 HOW TO SURVIVE FRIENDSHIP

A TALE OF TWO LUMINARIES

In July 1870, a boy named Nikola, the son of a Serbian priest, was living in eastern Europe, in an area that is now the Republic of Croatia.[1] Nikola had fallen ill in his teenage years, and doctors, having given up hope that he would survive, gave him the task of classifying and cataloging books at the public library. One set of books intrigued young Nikola more than any other, and according to him, contributed to his miraculous recovery. The books were written by US author Samuel Clemens, known better by his pen name, Mark Twain.

Years later, Nikola, now fully recovered from his illness and a budding inventor, moved to New York City, like many do, for the opportunity to fulfill his dreams. By 1890, he took a substantial step to fill those dreams by starting his own company: Tesla Electric Light and Manufacturing. The ill, young Croatian man was Nikola Tesla. In the years that followed, Tesla developed ideas, filed patents, and worked on bringing his inventions to reality through a series of collaborations with electric companies. This allowed him to see his designs of alternating current electricity systems, the very electric systems that are used in electric power cords and outlets around the globe today, come to fruition. At the same time, Clemens, who was a writer by trade but also a forward thinker with diverse interests, became inspired by Tesla's electric machines and saw them as the way of the future.

Because of their shared interest in electricity and innovation, Clemens, the father of US literature, and Tesla, one of the greatest scientific minds in history, started corresponding through letters, which then progressed to face-to-face meetings. When Clemens traveled back from Europe, where he was living and writing at the time, the two met in Tesla's laboratory. They also arranged meetings at New York's Player's Club, the first men's club in the United States. With this began a twenty-year friendship between two formidable forces of the late nineteenth century.

Not all of us have friendships like that of Clemens and Tesla. Our friendships are generally not newsworthy; nor do many of us get to play an inspirational role for the luminaries of our world. Still, most of us do have friendships, on average about four close ones, and for many of us our friends play an inspirational role in our own worlds.[2] These relationships are powerful; they shape us in important ways. Moreover, they are often enduring. Aside from sibling relationships, friendships can be the longest-lasting relationships of our lifetime.

GIRL SCOUTS, MONCHHICHIS, AND A MEMORABLE FOURTH GRADE SPELLING BEE

To understand the significance of friendships in our world, I'd like to start at the beginning. Try to think back to your first best friend. Who is this person? And what do you think drew you to that person more than anyone else?

Some of you might be thinking that it was chance or circumstance that attracted you to your first friend. To some extent, all friendships are determined by circumstances. Those related to your first best friend were due largely to your parents, like the country, city, and neighborhood you lived in, as well as your socioeconomic status, school, and affiliation with religious or other social organizations. But provided that your first best friend was not determined entirely by circumstance, like they lived next door or your parents were friends, it is likely that your first best friend was more of an intentional choice than you think. This is because from the time you are born, whether you know it or not, you are seeking out specific environments that match your genetic makeup through an *active (or selective) gene-environment interaction*, more commonly known as *niche picking*.[3]

In the early 1980s, when I was in second grade, my family moved from one little town in Indiana to another about thirty miles away. Despite it being so close, I knew no one, thus making it a tough transition, especially as we moved in the middle of the school year. Luckily, there was an existing infrastructure that helped me acclimate: the Girl Scouts.[4] The badge-awarding, cookie-selling, craft-making organization, which has now been in existence for more than a century, had everything seven-year-old me would have ever wanted in an extracurricular activity. The troop had after-school meetings, went on excursions, and even had an overnight Girl Scout camp. Moreover, this particular troop was really active. Our leader was the wife of the superintendent, and her daughter,

Julia, a perky, sweet girl with blond ringlets, was in my class at school. Entry to the Girl Scouts translated into invitations to Julia's house for her famed birthday parties, fun-filled adventures in her giant house where all the girls in attendance received gift bags stuffed with Barbies, paper dolls, and Monchhichis (Japanese stuffed monkeys that were popular toys in the United States in the early 1980s).

But Julia didn't end up being my best friend. Instead it was someone else in the Girl Scouts, a girl named Leigh Ann, who attracted my attention. When I consider what drew me to her, I think it was her calm, easy nature and sense of humor. She was always composed and seemed, in my eyes, to have it all together. Conveniently, she also lived in my neighborhood, but her house was far enough away that I wasn't able to ride my bike there alone until I was in fourth grade. That was the same year Leigh Ann beat me in our school's spelling bee. She and I were alone onstage for the heated last round in our school gymnasium. I misspelled *boundary* (the irony of which has never been lost on me, a relationship researcher), and she brought home the win. The next year, she and I went together, timidly, to the sixth grade class for advanced instruction in math and social studies. Leigh Ann and I were first and second in our class, interchangeably, throughout elementary school. She was, and always will be, my intellectual match, my best friend, and soul mate.

When I first met Leigh Ann, I didn't ask about her grades, and her calm demeanor didn't register in my brain as maturity. I was seven. At that time, I knew only that I liked her. Now I know that it was probably my genes that liked her. I was drawn to Leigh Ann, just like I've been drawn to books, spelling bees, and badge-awarding after-school enrichment activities. And she was drawn to me. We were fellow drops of oil in a glass of water. The only circumstance that pulled us together was that we were in the same glass. But the bonding that took place was a result of our genetic makeup. We were naturally attracted to each other. Alike in important ways. Perhaps, like oil in water, we had no choice but to bond. And just like if someone tried to separate the oil by mixing or stirring, it was difficult to separate us. And it's that *homophily*, our similarity, that has helped us, like Clemens and Tesla, sustain our friendship over forty years.[5]

But what value does a friend like Leigh Ann bring? A lot, according to research. The value of friendship was understudied in the early parts of the twentieth century. Prevailing psychologists of the time, like Sigmund Freud and John Bowlby, placed emphasis on caregivers along with the roles they played in shaping personality and attachment. In the latter part of the twentieth century, however, Erikson extended both the duration and net of influence in his stage theory of psychosocial development.[6] Erikson, a student of Freud's daughter, Anna, was in agreement with Freud and Bowlby that early attachments with caregivers and other family members shaped a child's personality.[7] He nonetheless departed sharply from Freud's beliefs about sexuality and libido as major underpinnings of human development. Erikson also held that as a person transitioned into adolescence and adulthood, relationships with individuals outside the family, like friends, peers, and workmates, became more critical to our socioemotional development. As an adolescent, for example, friends and peers shape our identity development in important ways that help us figure out who we are as well as what we want to be. Later, our friends and romantic partners are crucial in helping us resolve our crisis of intimacy versus isolation; like our caregivers did early in life, they teach us how to love.

Today, the value of friendships is no longer debatable or overlooked. Friends, especially close ones, are among our greatest sources of support, fulfillment, and intimacy. They are also the secret to a longer life. Remember the longitudinal Harvard study I mentioned in chapter 2? According to the current director of this study, the safekeeper of the data and designated interpreter of the study's meaningful contributions, "The clearest message that we get from this 75-year study is this: Good relationships keep us happier and healthier. Period."[8] Participants in this study with stronger friendships had longer, healthier lives, and when asked to give autobiographical sketches (detailing the important parts of their lives), more than one in five mentioned friends, along with family and career.[9] Science writer Lydia Denworth emphasized the value of friends as well in her 2020 book titled *Friendship*, which is subtitled "The Evolution, Biology, and Extraordinary Power of Life's Fundamental Bond."[10] The maintenance of strong friendships has been linked to everything from how well we sleep to our cardiovascular function. It also appears to be critical to longevity. In a 2010 meta-analysis of 148 studies including more than three hundred thousand people around the world, older adults with stronger friendships had a 50 percent greater chance of survival than those with weaker friendships, regardless of age or health status.[11] The most beneficial friendships, the ones that give you the greatest benefits in terms of mortality, are those that are richest in social support and reflect your engagement within a network. As you'll see in the next chapter, though, it is not necessarily the number or quality of the friendships you maintain that gives you these benefits; the diversity of your network matters a lot too.

THE VALUE OF TALKING TO STRANGERS

Never in history have we had more opportunity for diversity in our friendships than we do today. Of course, matching phenomena (whereby we are attracted to those who are similar to us in key ways) are still at play in shaping our preferences for our most important connections. Yet the internet opens the world to us, making it possible for us to develop meaningful peer relationships with people who come from vastly different backgrounds from all around the globe. In the 1960s and into the next few decades, the options for expanding our networks in this way were limited. Pen pals, the written snail mail correspondence initiative that matched individuals (including schoolchildren) to others in different cities worldwide, were one such option. In the 1980s, this extended to phone communication with the advent of party lines, the pay-per-minute 900 numbers that allowed you to chat with strangers for the exorbitant price of $0.45 to $1.00 per minute.[12] Although these pay-based party lines have lost their appeal among modern-day youths, this is not due to a lack of interest in talking to strangers. It is due instead to the fact that the internet has made these services free in the form of social media and other chat mediums, in which strangers can meet, congregate, and chat in dyads or groups.

People can meet in official "pen pal" online spaces to chat with people around the world and even engage in project-based learning together related to important topics, like climate change and robotics.[13] All on the internet, and all for free. Meanwhile, individuals today have nearly limitless options for communication with strangers, from gaming-linked chat applications like Discord to more traditional social media applications like Snapchat and the audio-based chat application Clubhouse. US singer and songwriter John Mayer even suggested a novel way to connect to others in a tweet from 2017: "They should let everyone on hold with customer service talk to one another."[14] Mayer's tweet actually gets to the heart of this book; the desire to expand our networks and talk with strangers reflects not only our attraction to the novel and diverse but also our fundamental need for human connection. And in a world wherein our daily pursuits for connection can get easily sidetracked to depersonalized technology, like with customer service calls, there is opportunity to develop positive relationships.

Like we sometimes do on airplanes. According to the study known as Flyland, funded by the British bank HSBC, more than half of airline passengers strike up a conversation with someone sitting near them, and 14 and 16 percent have made a close friendship or business connection, respectively.[15] One in fifty has also found a love match. These happenstance meetings have personal value in my own life. On one of my lengthy plane rides to Seoul, I spent the latter part of the twelve hours talking (and writing, for increased clarity) with a Korean businessman who had started a company, MyGenomeBox, that marries computer analytics with genomics to help people tailor their diets and exercise based on their genetic profile.[16] On a flight to Seattle, I met a former US Navy pilot who once had the job of landing planes on aircraft carriers and was now a distribution specialist for Amazon, leading its Prime Air initiative, which has just received Federal Aviation Administration approval for the use of unstaffed drones to deliver packages to US-based Amazon customers.[17] In 2020, I met a resident doctor flying from New York City, where he had been treating COVID-19 patients for eight straight months; he was finally going back home to visit his family in North Carolina. In my lifetime, I have met hundreds of people while in transit on planes and trains throughout the world. Some of them, like the three I mentioned, have inspired me, getting me to think beyond my own experiences to see the bigger scope of the world around me. Others have simply offered a kind word, conversation, or even just a smile that made my day a little bit brighter. Why did I strike up a conversation? Because I'm a happy person, and I want to be even happier.

According to University of Chicago business professors Nicholas Epley and Juliana Schroeder, "Feeling socially connected increases happiness and health, whereas feeling disconnected is depressing and unhealthy."[18] In their 2014 paper aptly titled "Mistakenly Seeking Solitude," Epley and Schroeder describe a series of studies that tested hypotheses related to inclinations toward and benefits gained from connecting with strangers. They hypothesized that people might be reticent to talk to strangers for a variety of reasons. It may be due to purposeful decision-making in the face of an overabundance of social options—a sort of social or cognitive economizing. Alternatively, it could be based on misguided assumptions that strangers would be poor conversational partners or inadequate sources of social support. It might also be due to social norms that deter engagement with unknown others.[19] Or perhaps, despite Aristotle's claims that humans are social animals by nature, we may have a biological tendency toward solitude. Primates, after all, have ancestral roots as a solitary species that only transitioned to pair and group social structures as activities switched from nocturnal (nighttime) to diurnal (daytime).[20] Whatever the reason, millions of commuters worldwide choose solitude, immersing themselves in their phones, books, or own thoughts, with nary a nod to nearby passengers. The next time you are on a bus or train, or in an elevator or waiting room, count the number of people you see who talk to a stranger. In some of our most public places, where we could most easily connect with and provide support to other humans, we choose to ignore them. As Epley and Schroeder showed, this could be to our detriment.

In their field experiments, research assistants approached individuals boarding morning commuter trains into downtown Chicago. They randomly assigned these commuters to one of three conditions. In the connection condition, they were told to try to have a conversation with a new person, attempt to find out something interesting about the stranger, and tell the stranger something about themselves. In the solitude condition, they were asked to sit with their own thoughts and focus on the day ahead. In the control condition, they were told to do what they normally do. Who do you think had the best experience? When the participants rated the positivity of their experience, those in the connection condition reported a significantly more positive experience than those in the solitude condition. Talking to strangers—even when forced to (recall the psychological reactance I spoke about in the introduction)—made people happy. This is not what people predicted would happen, however.

In a related study, commuters at the same station were approached by research assistants and asked to *imagine* they were in the three conditions that had actually been enacted in the first study. They were asked to predict how happy they would feel and how pleasant the experience would be in each condition. The commuters' predictions were in the exact opposite direction: people expected they would have the most positive experience in the solitude condition and the least positive experience in the connection condition. Epley and Schroeder replicated these findings in experiments targeting commuters on buses in Chicago too and found the same trends. Talking to strangers was a positive experience for people, whatever the mode of transport. But people mistakenly thought being alone would give them the most pleasure. Social engagement, it seems, is a key to happiness.

SOCIAL DEPRIVATION MIGHT ACTUALLY MAKE YOU DIE

Surely, though, not everyone had fun talking to a stranger on a train? Individual differences matter, and although the experience was enjoyable to the average person, some people likely hated it. Consider the introvert. Introverts would probably not get as much pleasure talking to a stranger as those who are extroverted. Or would they? Although this wasn't tested directly in the commuter studies, previous work has shown that introverts have as much fun acting extroverted as do extroverts.[21] The major difference between these individuals is their attitude, or how much fun they *think* they will have. Introverts, it seems, make significant *forecasting errors* and believe social events will be more negative than they actually are. These errors have consequences. Introverts will often skip out on social activities based on these faulty predictions, which may contribute to their characteristically lower levels of reported happiness as compared to extraverts.

If this isn't enough to convince the introverts among you to be more social, recall the meta-analysis of 148 studies I mentioned earlier in this chapter showing that good friendships can actually help you live longer.[22] Social isolation also emerged as an important predictor of mortality in that study. In this case, the more socially isolated you were, the more likely you were to die. Social integration therefore might actually be imperative to survival.

Fortunately, the world is full of online social networks. Perhaps it isn't really necessary to talk to a stranger or attend a party. Instead, maybe we can just hop onto social media from our own cozy beds, slipper and pajama clad, to get the social stimulation we need to live longer, happier lives? Unfortunately, the answer does not appear to be that simple. As mentioned, there are some studies, such as those involving US adolescents, suggesting that greater engagement with social media is related to worse mental health. Similar studies with adolescents in Iceland have shown the same trend.[23] But what about social isolation specifically? Perhaps social media can help people feel less isolated when they connect, even briefly, with others in online spaces?

This makes sense, but it doesn't appear to be the case. In a study involving young adults in the United States, those who reported the highest rates of social media use also felt the most socially isolated.[24] In fact, even after accounting for variables that might make people feel socially isolated, like age, sex, relationship status, and living situation, those in the top 25 percent of social media use were twice as likely to report feeling socially isolated as those in the bottom 25 percent. This is the paradox of social media. On the one hand, it opens up the world to us. On the other hand, it shrinks our lives to the confines of our four walls. Whether I am looking for support, connection, or even just entertainment, logging onto social media may not meet my needs. In fact, it may accentuate my loneliness. But we really cannot determine that based on this study. It was a correlational study, and it is impossible to determine the direction of causality. To figure out the direction of influence, we need to conduct an experiment.

So for the past few years, that's exactly what researchers have been doing. In 2018, University of Pennsylvania scientists recruited 143 undergrads for a three-week social media detoxing experiment.[25] When the students arrived at the lab, they were randomized into two groups. The first group was told to do what it normally would do regarding social media. Its use over the following three weeks would be tracked but not manipulated. The second group was told to limit its use of three platforms (Facebook, Snapchat, and Instagram) to only ten minutes each per day. Again, its use would be tracked to make sure group members were adhering to the protocol.

At the end of three weeks, people in both groups were less anxious, thereby suggesting that just attending to their social media use had a positive effect on anxiety. Those who detoxed, however, experienced even greater benefits. Those who limited their social media use to only thirty minutes per day were significantly less lonely and depressed than those who did not do the detox. Recall from the introductory chapter that more sex did not make married people happier. And recall that I pointed out a variety of issues in that study that made it unlikely for people to get happy throughout the experiment. Like the fact that they were being told what to do, and people don't like being told what to do. Yet in this experiment, where young adults were being told to limit their social media use (which was likely perceived as restrictive and undesirable to at least some), social media detoxing worked. It made people happier.

Like everything I discuss in this book, the answers are rarely straightforward and there are alternative interpretations to consider. It's important to keep in mind, for example, that this is a single study. Although the majority of recent studies show that excessive social media use has various negative effects, the findings are not consistent.[26] Additionally, in the study I outlined, researchers didn't explore whether *increasing* social media use would have a positive effect on mental health. A study like this would be much like the sex or doughnut studies. People would come to the lab, and they would be asked to increase their social media over a number of weeks or months. Then researchers would rate their levels of happiness, loneliness, and depression. This is unlikely to happen, though. Unless there are major shifts in the general sentiment of the research from "media exposure equals harm" to "media exposure equals benefit," it might be unethical to conduct this type of study. As psychologists, we aren't supposed to conduct studies in which the potential harm outweighs the benefits. This is likely a dead-end pursuit. But there are many other issues to consider too. Psychologist Ethan Kross and colleagues describe several problems with the social media research to date.[27] For instance, the overall negative effects of social media on mental health are small and usually not clinically significant. Moreover, people use many different terms (e.g., social media use, social networking, or online interactions) and measure usage in different ways (e.g., Facebook, Snapchat, or all social media), thereby culminating in a *jingle-jangle problem*. Too much diversity produces too few clear answers.

Finally, and I think this is my primary criticism of this line of research, the decreases in anxiety and depression that these young adults experienced as a result of the social media detox may not be due to the fact that they were using social media *less*. Rather, the decreases in anxiety and depression might be because they were doing other activities *more*. Indeed, time spent on technology displaces time that could be spent doing something else. And it may be that these alternative activities are more beneficial to health and wellness. For example, perhaps people chose to exercise for thirty minutes instead of engaging with social media. It is well documented that exercise helps to diminish symptoms of depression.[28] So perhaps the same or greater benefits would have been realized if instead of decreasing social media use for thirty minutes each day, the researchers had those in the experimental group increase their exercise by thirty minutes per day. Alternatively, they could have had them complete a task, play with a pet, volunteer, or get a massage. There are lots of ways to lower levels of anxiety and depression. The important thing to note is that what you take away may not be as important as what you add.

FRIENDS ARE LIKE MONEY

All of this said, there is no debate about the benefits of social engagement generally. Experts are in agreement that social engagement is beneficial, and our friends help us live longer, happier lives. Friends might also make us smarter. Not only can they bring diverse perspectives and experiences into our worlds, but they might affect our cognitive growth in more direct ways by, for instance, recommending books, music, art, and travel. Moreover, simply maintaining friendships requires cognitive effort. In order to be a good friend, we must track those in our social circle in both time and space. "Where is he going on Friday night?" "Is she finished with work?" "I haven't heard from her in a while, I should probably send a message." And we must also be attuned to stimuli that help us understand our relationship to them and their relationship to others.[29] A new dating partner on the scene, for example, might mean that you will skip your regular Sunday brunch. These simple acts of accommodation are actually sophisticated ways of thinking about complex human relationships. And this is one of the keys, posits Robert Dunbar, a British anthropologist who developed the *social brain hypothesis*, to human evolution. According to Dunbar, forming social groups helped us grow larger and more complex brains.[30]

Friendship might also be rewarding from a biological perspective. If you've never heard of the *mesolimbic pathway*, please let me entertain you with a brief and simple lesson in brain biology. The mesolimbic pathway is the set of neurons that connect the ventral tagmental area (a part of the midbrain that is known for its concentration of neurons that produce dopamine) to the ventral striatum (a part of the basal ganglia in the forebrain that contains the nucleus accumbens and is known for its activation during reward anticipation).[31] *Dopamine* is the neurotransmitter associated with, among other things, pleasure. Although there are various brain areas involved in the experience as well as interpretation of reward and pleasure, the mesolimbic pathway—with its large concentrations of dopamine—is often referred to as the *reward pathway*. When we have a pleasurable experience, dopamine travels within neurons along this pathway, and this is what triggers our positive feelings. Hence both the mesolimbic pathway and dopamine have been the subject of much scientific inquiry regarding addiction to everything from cocaine to chocolate.

But how does this relate to social engagement? First, a bit more on rewarding experiences. There are two types of *reinforcers*, or stimuli that increase the rate or likelihood of a behavior. *Primary reinforcers* are biologically determined. They are stimuli that our bodies respond to naturally without any training or experience. Sex, food, positive touch, and drinks—these activate the dopaminergic pathways in our brain, like the mesolimbic system, and we experience pleasure. We don't have to learn about these; our bodies just naturally respond. *Secondary reinforcers*, on the other hand, are not biologically programmed. Reinforcers like money, applause, or grades are only reinforcing because we associate them with primary rein-forcers. Money can buy food and drink. In some parts of the world, money can even buy you touch and sex. Through the pairing of these pleasurable experiences with money, we come to appreciate money. Money becomes rewarding too.

Friends are like money. On their own, they do not pro-duce any type of natural biological reaction. But when we learn to associate our friends with primary reinforcers, like food, drinks, and hugs, then the friends become the reward. It makes sense, then, that we normally spend time with friends at dinners and parties, while doing other things that we enjoy. Essentially, we glean rushes of dopamine and oxytocin through the sensational experiences, but may attribute those biological rewards to our company. In time, with greater attachment bonds, our brain adapts from the simple reward motivations associated with novel stimuli to more attachment-oriented motivations associated with long-term bonds.[32] This is one of the reasons that social connection is believed to help make us more resilient to addiction. Perhaps this is why we always hear stories of people eating ice cream after a breakup. Deprived of the oxytocin elicited in the company of those we love, we crave the simple dopaminergic pleasure of a pint of Ben and Jerry's.

HOW MUCH ICE CREAM DOES IT TAKE TO FILL AN EMPTY HEART?

But what if this deprivation is not a momentary experience? What about those people who struggle to maintain social relationships? How much ice cream does it take to fill an empty heart?

In a study using the Gallup World Poll data, more than half of the unhappiest people in the world reported that they had no social support.[33] They are also stressed, sometimes in pain or poor health, and don't have the resources to meet their needs. Luckily, these unhappy people comprise only a small percentage of the overall population—0.6 percent. But they are not the only ones reporting difficulty with social support. The moderately happy people, approximately 8.6 percent of the world's population, report lower levels of social support than those who are in the happiest group.

Yet social support isn't the only measure reflective of an empty heart. Surveys of loneliness across the United States, United Kingdom, and Japan show that millions of people are lonely worldwide (22, 23, and 9 percent, in these countries, respectively). As with a lack of social support, loneliness is more likely among certain groups of people. In this case, those who are lonely are more likely to be in the younger age groups (under fifty years of age), unmarried, in poor health (both physical and psychological), and in more dire financial circumstances.[34]

Loneliness and lack of social support are significant public health risks. They are also costly to humanity. Out of the wells of loneliness sometimes spring addiction to drugs, depression, and suicide, which come with significant costs to both human life and our world economy. The total economic burden for these issues totals hundreds of billions of dollars in the United States alone.[35] Loneliness is such a prominent and critical threat that organizations have set forth missions to tackle the growing problem, such as the Cost of Loneliness Project, headed by renowned health strategist Lucy Rose and numerous researchers associated with the American Psychological Association.[36] These initiatives exemplify a new age of recognizing loneliness as an extremely detrimental cost to human function. It shows that we are no longer ignoring the psychological ails of humanity. This is promising. The only way to combat this epidemic is through concerted efforts that recognize the weight of the issue.

Other people are taking a more practical approach. Recognizing the value of friendship, businesses worldwide are offering online friend-finding services. The popular dating site Bumble has one such offering, Bumble BFF, which is advertised as offering "a simplified way to create meaningful friendships." Instead of perusing profiles for potential dates, individuals on Bumble BFF try to find friends with whom to connect. Although one user reported finding more club promoters than actual friend prospects, I suppose, like online dating, the idea of swiping right for a BFF might gain some traction.[37] It doesn't seem to be the answer just yet, however.

If that's unsuccessful, you can also rent a friend. In places like Japan, these rent-a-friend services have long been in place, with businesses such as Family Romance offering hundreds of actors, from children to seniors, to pose as anything from love interests to parents to mourners at a funeral.[38] As international websites like RentAFriend popping up, though, others around the world can rent the services of a local friend for $24.95 per month plus the cost of your "friend's" time. Perhaps you are looking for someone to go to a movie or dinner with, or you'd like a companion to travel or work out with. With over six hundred thousand friends on RentAFriend's website, there appear to be lots of options. But how fulfilling are these rented friendships? And can these online services lead to real friendships? Friendships, it seems, are built on mutual trust, disclosure, and similarity.[39] These are the building blocks of intimate friendships. Unfortunately, the friend you swiped for because of a mutual interest in yoga or rented to go to the movies may not check all these boxes. That said, according to one renter, who engaged in renting a friend for a news story, it "felt better than being lonely."[40]

There's a trick of the mind that we like to play on ourselves. It's called the *just world phenomenon*. We like to think that the world is fair, good things happen to good people, and bad things happen to bad people. Thinking in this way helps us cope with a world that is often drastically unfair. These lonely people, the ones the Beatles wrote about in the song "Eleanor Rigby," didn't do anything wrong to end up where they are. They may have always been down and out in ways that biased the world against them. And just like we all can't be friends with the world's shining luminaries, not all of us have the privilege of having good friends to buoy us through rough waters. We nevertheless all deserve the benefit of friendship. And the more the world opens through the ties of the internet, the more opportunity people have to forge connections with people with similar interests and experiences. Even if this is the sole benefit of our connected world, it's enough.

HOW TO SURVIVE FRIENDSHIP

A Tale of Two Luminaries

Survival tip #14   If Clemens and Tesla could forge and maintain a friendship in the early 1900s, while living on different continents and without the aid of transatlantic flights, you have absolutely no excuse for not keeping in touch with your friends. If there is no reciprocity or they don't inspire you, find new ones. But if you already have friends, and they are of even minimal value to you, make the effort to sustain these important relationships. It took at least five days for Clemens to cross the Atlantic to visit Tesla. You can take twenty minutes to call your best friend.

Girl Scouts, Monchhichis, and a Memorable Fourth Grade Spelling Bee

Survival tip #15   Drift naturally to those to whom you are biologically drawn. Find them in coffee shops, or at concerts, sporting events, or parties. You can even find them online. You choose your environments. Trust your instincts about where to wander, and while there, look for your soul mates.

The Value of Talking to Strangers

Survival tip #16   Overcome your reluctance to engage with strangers. Connect with them during commutes or solo adventures. Ask them an interesting question like, "How would your best friends describe you?" Minimally, you may have a pleasant experience and learn something about one of our world's beautiful humans. Perhaps you enrich their day. But in the best case, you forge a meaningful and enriching connection that opens your mind and heart.

Social Deprivation Might Actually Make You Die

Survival tip #17   I admit this is a bit dramatic, but social isolation is bad for humans. We are built to be social beings. Think about Tom Hanks's character in the movie *Castaway*.

He was so socially starved, he started taking to a volleyball. Even the robots in the movie *I, Robot* huddled together in their metal storage bins. Sure, these are the musings of Hollywood writers, but they are reflective of basic human needs. Find connection wherever you can, and if that happens to be online, go for it. Just also make sure that you are balancing your life with other activities that help elevate your mood such as exercise.

Friends Are Like Money

Survival tip #18   I'm not sure anyone needs additional incentive for developing friendships, but having friends, aside from building your social capital, is actually good for your brain. Spending time with friends can have a similar positive effect on the brain as chocolate or ice cream, but unlike these tasty treats, friends are calorie free. One way to strengthen your brain's positive reactions to friends is to aim to pair friendly outings with other types of rewarding experiences like exercising and dining out. Eventually, the familiar can bring as much pleasure as the novel.

How Much Ice Cream Does It Take to Fill an Empty Heart?

Survival tip #19   There are people in this world who are lonely. Truly and deeply lonely. We can no longer ignore their suffering. Dismiss your false assumptions about what they did wrong to end up in a place without social ties, and offer them your sympathy and support. In the two years before my grandmother died, I called her almost every time I commuted to work. She was blind, in a wheelchair, and spent the last of her days on this earth in a care facility where the managers once tried to pawn off construction paper snowmen building as a group leisure activity. Sometimes I caught her. Sometimes

I didn't. But I will never regret how I spent that time. Use your cell phone to connect. Aka call your grandma.

Bonus Technology Tip

Every time you're waiting in a line or commuting, consider sending text or voice message to a few friends. Tell them you're thinking of them. Mention what you cherish about them. Ask them about their lives. It might help you maintain connections and deepen relationships, but it also might help you maintain a healthy life balance, where your consumption matches your output. Activities like reading, watching television, and listening to music are consumption, while activities such as exercise, making meals, and sending messages are output. Be intentional about seeking balance.

4 HOW TO SURVIVE THE INTERNET

THE SOCIOTECHNICAL PANOPTICON

In the eighteenth century, Jeremy Bentham, an English philosopher, and his brother, Samuel, a trained shipwright and engineer, developed the idea of the panopticon.[1] Samuel was living in Russia at the time, working for Prince Potemkin of Krichev. Charged with overseeing shipbuilding and various manufacturing ventures, Samuel conceived of a unique solution for the challenge his supervisors faced: how to train and monitor workers who were inexperienced in their craft. Through a series of letters and conversations, the brothers refined the design of the panopticon, a structure that would allow a single person in a middle tower to oversee the doings of the individuals in various rooms around the tower, built as a tiered rotunda. In England, Jeremy furthered the design, and with his interest in social reform, applied the panopticon structure to a prison, wherein a central tower would be surrounded by all the prisoners' cells. All the cells would be open to the watchful eye of the guard, but because of the panopticon's design, none of the inmates would know if they were being watched. A key feature of the design was its potential effect on those being watched: it was assumed that because inmates would not be able to discern whether or not they were being surveilled, they would regulate their own behavior. Panopticism therefore emerged as both an architectural design that would be helpful to those wanting to observe others' behavior and a theoretical construct of surveillance.

It was not long after social media emerged that individuals began to apply the concept of the panopticon to our sociotechnical world.[2] The design is eerily consistent with the architecture of social media, but with social media, the panopticon design is also reversed: the person stands alone in the middle of the structure with observers in the cells surrounding them, watching their every move. Whether the person is actually being watched is irrelevant; all that really matters is that the person perceives they are being watched. Still, in the case of the internet and social media, perception matches reality.

The amount of data being gathered today by our various devices is staggering. It is a market so vast in breadth and depth, and a practice so controversial, that teams of scholars in the fields of philosophy, data ethics, social science, business, and computational science have been studying these issues for decades. Although in the early days of social media the extent of the surveillance was likely unknown by users, who quickly checked through privacy policies while opting into online social networks, numerous events in the early twenty-first century brought surveillance issues into the arena of general public discussion. This includes Edward Snowden's 2013 whistleblowing on the US National Security Administration's worldwide surveillance and the 2018 exposure of

Cambridge Analytica, wherein Facebook allowed the political consulting firm access to the data of approximately eighty-seven million users.[3] These key events also spurred support for large-scale legislative changes in data privacy and security such as the General Data Protection Regulation, enacted by the European Union in 2018 to regulate the ways in which its citizens' data could be used by companies worldwide.[4] While I'll defer to experts on data privacy and big data to explain the technical aspects of how and why user data are being collected as well as productized, I want to give insight into my personal experiences with the internet guards, and how they factor into our relationships with proximal and distal connections.

Since 2015, I have served as an expert witness and consultant for criminal courts for a variety of crimes, usually related to sexual assault, intimate partner violence, or cybercrime. In this role, I have had the opportunity to review thousands and thousands of bits of digital evidence. Saved and deleted images. Parts of text conversations individuals provided to authorities and parts of text conversations they did not provide to authorities that were later recovered with sophisticated phone extraction tools. Conversations that occurred over an evening and those that took place over a year. There is nothing quite like reviewing evidence from a criminal investigation to make you fully aware of the vast amounts of data being collected on your devices. I have seen the "private" threats perpetrators have sent to those they victimized. I have read through thousands of text messages documenting the stages of a sadomasochistic relationship, from the "are you interested?" stage to solidified plans for BDSM scenes.

I have witnessed the compliments and pleas cheating partners bestow on their paramours. Through these experiences, I have come to view these digital records as the most intimate details of our existence, revealing the backstage of our lives, far removed from the identities we carefully curate for public display. In a world dominated by computer-mediated communication, our phones and computers—now sophisticated tools of surveillance—capture "all of us." Your phone is essentially a pocket-size panopticon.

Modern philosophers, such as professor of the philosophy of technology Alberto Romele and Italian colleagues, have taken this somewhat dystopic proposition a step further, suggesting that those of us who elect to participate in our sociotechnical world are submitting ourselves to "voluntary servitude," wherein we willingly disclose our personal details online despite the fact that we now know these data are being collected and used by the social media sites and technologies to which we subscribe.[5] To these surveillance entities masquerading as social tools, we are data points that can be studied, monetized, and manipulated to perform various actions. From buying products to voting for political candidates, we are cogs in a giant marketing machine. Our submission to these systems, through our continued participation, exemplifies our denial, which according to sociologist Stanley Cohen is a perpetual state for humans in a world full of atrocities.[6] As thoughtful observers, we acknowledge the ills of the world, but in order to cope with the uncomfortable feelings these acknowledgments provoke, we rely on Freudian defense mechanisms like *denial* to intentionally dispel these ideas from conscious thought.

I <3 MY PHONE

In this light, submitting to these internet entities must seem a foolish misstep. Yet as I've written this chapter, I've checked my Facebook, Instagram, Twitter, and LinkedIn feeds at least a hundred times. Why? Stated simply, I and the 53 percent of the adults across the world who use social media must believe that the affordances of these mediums are more beneficial than the costs.[7] Stated not so simply, my relationship with social media is . . . complicated.

My declaration concerning my relationship status with social media is bold in that I'm acknowledging we are in a relationship, and I'll go one step further and state it's actually a relationship with my phone. One might argue that these technologies are simply the mechanisms through which we develop and sustain relationships, and it is impossible to have a relationship with an amorphous or technological entity. Yet in many ways, my phone has taken a central relationship role, as it has in the lives of countless other cell phone users across the globe. How key is this role? A 2016 online data-tracking study by dscout, an online qualitative research platform, showed that the average person (randomly selected from a pool of a hundred thousand) touched their phone 2,617 times per day, and the heaviest users (the top 10 percent) touched their phones a staggering 5,427 times daily.[8] By these estimates, I am likely no more than an average user, and yet my phone is certainly a central relationship figure in my life.

Through its lights, sounds, and vibrations, my phone makes bids for attention, and I respond. Much like the way I respond to others in my life who make these bids (e.g., my husband and children), I turn to it, attend to it, and seek to resolve the issue that prompted the alert. My phone is probably the most demanding entity in my current world. As a developmental psychologist, I have taught my students that responsiveness is one of the key elements of parenting and one of the most impactful things you can do as a parent to nurture a child. Hence through my responsiveness to my phone's bids, I have nurtured it as well. But it's not only responsiveness that's solidified our relationship. I carefully wipe its screen to remove smudges (social grooming). I carry it with me everywhere I go in either my purse, hand, or pocket (skin-to-screen bonding). I get nervous if I cannot find it (separation anxiety). We are bonded, and I am smitten.

This relationship has not gone unnoticed by others in my world. And if you are anything like me, it has not gone unnoticed by others in your world either. Along with family scientist Brandon McDaniel, I have been exploring the ways in which technology is interfering in dyadic relationships via the little, everyday interruptions in our interactions termed *technoference* or *phubbing*.[9] Since 2016, McDaniel and I, along with other researchers around the world, have found some consistent trends.[10] Specifically, people sometimes choose to interact with their phones over the human others in their world, and this can cause conflict and jealousy in couple, family, and friend relationships. In turn, this conflict and jealousy is related to lower levels of relationship satisfaction, and it also compromises intimacy. Unfortunately, this technology interference is affecting some of us almost every day. In our 2019 study on the topic, McDaniel and I employed a daily diary study, where we asked both members of a romantic couple to chart the technoference they experienced and their feelings every day over fourteen days.[11] The findings were striking. Most couples (72 percent) reported technoference in their interactions with their partner over the course of the two weeks. More important, on the days that participants reported more technoference, they also reported more conflict over technology, less positive face-to-face interactions with their partner, and more negativity regarding their moods and feelings about their relationships. These results emerged even after controlling for the amount of time spent with a partner—meaning that it wasn't simply that time spent with the technology was displacing time spent with the partner (and contributing to more negative consequences). Instead, it appears that the interruptions themselves had a negative effect. Or rather, when people made a choice to interact with their phones over the people in their world, it had a profound negative effect.

Why might we feel so rebuffed when a partner or friend chooses to interact with a phone as opposed to us? According to the theory of *symbolic interactionism*, our interactions with others are laced with messages, and those messages help us determine our role in that person's life.[12] When a person elects to attend to their phone rather than to us, especially when we are making efforts to engage them, it sends the symbol that the phone is more important than we are. Even if this is only a momentary experience, it can feel like rejection. It might violate social expectations about attentiveness and responsiveness too.[13] For either or both of these reasons, when a conversation partner attends to their phone instead of us, it might register as a relationship cost. According to the *social exchange theory*, our decision to stay in a relationship involves a constant evaluation of the costs and benefits of that relationship.[14] Essentially, we are keeping tally of pluses and minuses for our partners, and to stay invested and committed, a balance must be struck.

Regarding my relationship with my phone, the balance always tips in its favor.

Certainly, there are costs to this relationship. Aside from the monetary costs and potential for technoference that arise from my commitment to it, my phone brings a host of other costs. Most notably, it is my biggest distraction from my work, family, and friends. Regardless of where I am, when an email or text pops up, I feel compelled to check it. I also fall into rabbit holes of inquiry that start with reading a simple article about Fyodor Dostoyevsky's conceptualization of love and end two hours later having read about the definition of love according to twenty different philosophers. Thanks to the documentary *The Social Dilemma* and other recent commentaries on the tech industry, I now understand that these compulsions are rooted in purposeful design.[15] Although I understand *why* I fall prey, I still recognize myself as prey, however, and that leaves me discontented.

From a larger, societal standpoint, phones and technology use may also be causing discontent. In the early twenty-first century, there has been much media attention directed toward the research of US psychologist Jean Twenge and colleagues. Their studies have demonstrated a rise in rates of depression and anxiety corresponding to increased rates of technology usage among US adolescents and young adults over the past decade.[16] Suddenly, everyone is concerned about technology use as well as how it might be contributing to the degradation of social relationships and individual mental health. The argument goes something like this. Technology helps us form relationships, certainly. Yet now everyone is sitting in their bedroom, on their phones and computers, and by connecting with others online, they are missing out on the face-to-face interactions that help keep us happy and feeling socially connected. Even worse, going online and on social media is making us stressed, lonely, and depressed, particularly when it results in problematic and compulsive use.[17] Although statistically sophisticated analyses show that the overall negative effects of technology use on adolescents' psychological well-being are small, and connecting with people online can help reduce loneliness, this has not abated concerns.[18]

THE SOCIAL NETWORK

Nonetheless, most of the people in the world and I have made decisions to keep our relationship with our phones, likely due to a cost-benefit analysis where the phone won. Why does the phone always win? Simple: it helps us fill our needs for intimacy. Instantly. Phones allow us to connect with an unlimited number of people, 24/7, at the touch of a button. The world is literally at our fingertips. It is a cornucopia of potential social pleasures. We can connect with people we know. We can connect with people we don't know. And the more types of people we engage with, the better off we are.

There is a measure aptly titled the *Social Network Index* (SNI) that captures the diversity of our social connections. It was first introduced in 1979 as the Berkman-Syme SNI, which measured the number of connections individuals had to close friends, relatives, group-based contacts (e.g., social or religious), and others with whom they could get advice, support, and affection.[19] The SNI measure was expanded in the 1990s, however, by psychologist Sheldon Cohen at Carnegie Mellon and colleagues who were investigating the relationship between the number of social ties someone had and their susceptibility to the common cold.[20] Their actual measure is somewhat more complex than what I present below, but it is the gist of the exercise. [21]

In the spaces below, place a tally mark for each person in the following categories who you interacted with (on the phone or face-to-face) at least once in the last two weeks:

___ Spouse/partner ___ Parents ___ Parents-in-law

___ Children ___ Other close family members

___ Close neighbors ___ Friends ___ Workmates

___ Schoolmates ___ Fellow volunteers

___ Members of groups without religious affiliations

___ Members of religious groups

Now to compute your SNI, give yourself a point for each *category* in which you have at least one tally mark. If your SNI is one to three, your social diversity is low; if it is four to five, you have a moderately diverse network; and if it is six or greater, you have a highly diverse network. Now add in the contacts that you communicated with via text. Did your number increase? If so, you've illustrated the widening force of the internet. And you just might have improved your resilience.

According to Cohen and colleagues, having a highly diverse network makes you more resilient; in their study, those with six or more connections were four times less susceptible to common colds than those with one to three social connections. It is important to note here that is not the total number of connections or your extraversion that matters (these were controlled for in the study); instead, it is the diversity of the network. Phones allow us more diversity. Rather than communicating with only those in our homes or workplaces, we can connect to distant others and bring diversity to our social experience. Moreover, researchers have long purported that phones and social media facilitate the acquisition and maintenance of *social capital*, or the collection of resources we garner from our relationships.[22] The acquisition of social capital helps us feel good about ourselves in the "I was able to make this friend/connection and therefore I must be good" kind of way. Although recent studies with US adolescents have shown that the more online social capital you have, the more likely you are going to experience stress when exposed to online risk (e.g., information breaches and exposure to explicit content), this might be an "all eggs in one basket" phenomenon; social capital distributed across online and off-line contexts would likely offer a more protective effect.[23]

THE SOCIAL SUPPORT NETWORK

But it is not only the diversity of our social network or acquisition of social capital that gives us resilience and esteem.

Once someone makes it into our social network, they can personally give us social support. The term *social support* is now part of our everyday vernacular, but it wasn't until the 1980s that researchers across different disciplines began to study intensely what social support is and the benefits it provides.[24] What is clear from this work is that social support is extremely beneficial to us across a wide array of situations; it has been linked to better outcomes in everything from mortality and recovering from cancer, to stress and coping.[25] And although the definition of social support varies somewhat across disciplines, many generally accept that it can be divided into four main categories of support that we receive from others:[26]

Emotional: encouragement, concern, care, and sympathy
Instrumental: tangible help
Informational: information, advice, and feedback
Appraisal: feedback that helps with self-evaluation and goal setting

It's easy to see how our phones are bridges for emotional support. In times of stress, triumph, or even boredom, we can use our phones to connect with friends and family members. Additionally (or alternatively), we can disclose our woes and successes with our friends and followers in more public ways via posts on social media. When we elect to disclose on social media, we have two options: share big or small. It is not the subject or scope of the disclosure that determines the category. Rather, I have come to consider "sharing big" as anything that reveals individuals' deepest thoughts and concerns, and "sharing small" as the watered-down posts we make when we want to reveal some bit of information, but not ourselves.

According to the dual process model proposed by Stanford researchers Mufan Luo and Jeffrey Hancock, these social media self-disclosures, regardless of their width or valence, both affect and are affected by psychological well-being, and can result in significant benefit or distress through perceived or enacted social support.[27] So why might people share experiences like their pet's death on social media? Because it can have a profound effect on psychological well-being. Each of the likes, cares, hearts, and comments give a little hit of dopamine, which makes us feel good in the moment and also provides reinforcement for future social media interactions.[28] Moreover, the totality of these symbols is a physical manifestation of a village of sympathetic concern, which can make us feel supported while simultaneously meeting our basic needs for belongingness. Phones and social media offer instant access to this village.

But emotional support is not the only benefit. We use our phones to get actual, tangible support, like getting a job, help with a move, or a charitable donation. The benefits of phones can extend to information seeking too. Through our phones (and other connected devices), we have access to a wellspring of information. For those of us who grew up in an era when the primary information sources in the home were a stack of encyclopedias, a dictionary, and back issues of *National Geographic*, this information source benefit cannot be understated. We can find instant answers to almost any question we could ever invent. Even if the information secured is not of immediate value, information seeking itself is rewarding. In a classic study, neuroscientists Irving Biederman and Edward Vessel exposed humans to novel stimuli using pictures. They found that the images that were most novel and associated with other stored information in the brain stimulated the most mu opioid receptor activity, which is linked with increased pleasure. The process of discovery is so rewarding from a neurological perspective that we actually *crave* information. According to these scientists, we are "designed to be infovores," and the acquisition of knowledge provides an evolutionary advantage for both an individual (related to greater mate selection) and our species.[29]

Finally, the appraisal dimension of social support is probably the most controversial, at least as it pertains to social media. Through our social networks, we gather information about ourselves. How much are we liked? Do others care about what we say? How do our lives compare to those of others? Suddenly we have become the guards at the center of the panopticon.

TIDYING UP YOUR BRAIN

As humans, we spend a good amount of time thinking about what other people think. Although this is not meant to be a controversial proposition, I am sure that some of you are saying to yourselves, "No, I don't do that," or "I don't care what other people think." This is understandable. Any opposition you might feel to this statement might stem from well-meaning advice givers, who in effort to direct you away from self-criticism and anxiety emerging from others' evaluations, have assured you that what other people think of you does not matter. To that I will say simply, "Yes, it does," and urge you to read on as I explain why.

I'll begin at a basic level of human interaction with a statement we should all be able to endorse easily: when we act, it is likely and appropriate that we would think about how our actions affect others (e.g., "What does X think about what I just said?"). Still, our thinking about others' thoughts can also get more complex. We can widen the scope of inquiry by focusing on what someone thinks about the way we feel about someone else (e.g., "What does Y think I feel about X, considering what I just said?" or even, "What does Y think X feels about me, considering what I just said?"). These are not necessarily negative or maladaptive thoughts. Quite to the contrary, these are complex ways of thinking, employed by the most socially skilled among us, to help us develop and maintain social connections.

Although important, this way of thinking is not inborn; it is a sophisticated and critical skill we hone over our lifetimes. As toddlers, we don't consider our effects on other people. We make demands, cry, and maybe even throw tantrums in an egocentric way, wanting only to satisfy our own needs and not caring about anyone else. It is only as we learn more about the world, when we develop something called *theory of mind* between the ages of about three to five, that we begin to recognize that other people have thoughts and feelings, and that our actions might affect them. The theory of mind, by definition, is an ability to understand that both you and others have mental states, and that others' mental states might be different from your own and even different from what is reality. This understanding of one's own mental states gives individuals great power because once you realize you *have* thoughts and feelings, you can learn to *control* those thoughts and feelings. One of my favorite stories about the power that this skill affords comes from one of my university colleagues and a conversation he had with his daughter while driving her to preschool:

> "Dad, do you know what I'm thinking about?"
>> "No, Claire, what are you thinking about?"
>> "I'm thinking about Christmas."
>> "Very nice, Claire."

> A few minutes later . . .

> "You know what I'm thinking about now, dad?"
>> "No, Claire, what are you thinking about?"
>> "Now, I'm thinking about Halloween," Claire said with a smile.

As a preschooler, Claire not only showed her dad that she had theory of mind but also demonstrated an impressive mastery of directing those thoughts to control her own internal world. I contend that this is the skill that the advice givers are trying to get you to develop. Essentially, they are not saying, "Don't think about what other people think," which would be almost impossible for those of us over the age of five who have even a bit of empathy. Rather, they are saying, "When you think about what others think, use it for its potential good and then dispose of it," in a Marie Kondo kind of way.[30]

Dear thought: "You served me well, now goodbye."

Considering the average human has more than six thousand thoughts per day, this type of cognitive purging would likely do us good, freeing up our mental resources for more positive thinking.[31] But just like ridding our homes of unwanted items is a challenge for many of us, so is ridding our brains of unwanted thoughts and feelings. Particularly for those prone to anxiety and depression, worry and *rumination*—or stewing on negative thoughts associated with past events—is common and maladaptive.[32] Worry and rumination are also predicted by verbal intelligence, meaning that the smarter we are (in a verbal sense), the more likely we are to have a ruminating and worrying mind.[33] This is why many of us find it difficult to shift our cognitive energy away from the past events that vex us and toward the things that give us joy. Adding more fuel to embers of worry and rumination, the current landscape of technology-mediated social communication makes it especially difficult because of the mostly permanent nature of our digital steps.

Destroy after Reading

Thirty years ago, when we self-evaluated our effects on others, we relied on imperfect and quickly fading memories of events to help fill gaps in our perceptions. Luckily, the mind can be forgiving to the self, forgetting some damning details of past events and remembering the most promising via a pronounced *positivity bias*—a trick of memory that allows us to maintain our self-esteem, ego, and positive outlook.[34] This tendency to better remember positive details over negative ones might be particularly pronounced in the Pollyannas among us. According to the *Pollyanna principle*, people in general have a tendency to recall more pleasant than unpleasant details.[35] And those of us who do this the most also rate ourselves highly on happiness and optimism.[36] So happy, optimistic people tend to remember happy, optimistic details. But what about when they don't have to rely solely on their memory? What happens then?

Unfortunately, today's world of computer-mediated interactions is much less forgiving, and optimism can't help you erase black-and-white realities. Social interactions are often recorded permanently in the form of messages and pictures that one can go back to view and review repeatedly. In contrast to face-to-face interactions and phone calls, many digital communication channels (e.g., text message and email) are rated highly on a measure called *persistence*, meaning that the communication that occurs over these channels is perceived to be relatively permanent.[37]

How many times have you looked over an email or text message you've sent, rereading it to see if you said what you wanted to say in the way you wanted to say it? Ever sent a follow-up message correcting yourself? Clearly, some people stew on and regret messages they've sent. Luckily, ephemeral messaging applications, like Snapchat, have arisen on the scene too, providing us some reprieve from the persistence of our digital steps by employing a "destroy after reading" design. This design allows users to dictate how long a viewer has access to a message, which provided the viewer complies and does not screenshot or save a message, erases the message within the application. Facebook and Instagram's "vanish mode" and WhatsApp's "disappearing messages" updates in 2020 supply a similar experience.[38] Some messaging applications, like WhatsApp, even have a high degree of editability, giving users the option to erase sent messages if they change their minds soon enough (e.g., within the hour). In short, there are ways for you to avoid the permanence of digital steps in your private communication, and there are mediums that allow you to do exactly that.

Letting go of others' thoughts and our own online missteps might be especially difficult if our thoughts concern our close social connections, which are essential to our feelings of belongingness and love. This too is a natural human inclination, which can be best understood within two larger theoretical frameworks. First is Albert Bandura's social learning theory, which suggests that we do not become who we are in a vacuum.[39] Rather, we are influenced by those around us. From the time we are babies, we watch and mimic others. We learn vicariously from the ways that others are rewarded and punished, and we adapt our behaviors so that we can maximize rewards and avoid punishment. Hence our social contexts matter, and approval and disapproval from others shapes our future behavior. It is a basic tenet of human learning. Meanwhile, the importance we attach to this approval is related to another theory, Abraham Maslow's hierarchy of needs, a pyramid reflecting human motivation toward an actualized self, with rungs depicting basic human needs.[40] Right above physiological needs, like food and water, and safety needs, like a stable job and place to live, humans have a need for love. As children, these needs are often satisfied by a small number of people, usually family members, but as we spend more time in public spaces with peers and other agents of social influence, our need for love and belongingness expands to fill these spaces. Framed in accordance with this theory, caring about what people think is a protective process, designed to help us build a net of belongingness, which includes even distal connections, and safeguard us against the loss of love from those with whom we share our closest bonds.

Assuming that you've now embraced the idea that you do think of others' thoughts and feelings, and that is a good thing, I want you to direct your attention toward the people whose thoughts consume most of your time, in a relative sense. If you go back to your convoy diagram from the introductory chapter, some of the same people will probably come to mind. Yet the amount of time you spend thinking about another's thoughts might be inversely related to their distance from you in the diagram. You might also spend time thinking about the thoughts of people who didn't make it onto your diagram at all. Why might this be?

Uncertainty reduction theory, which suggests that individuals have a need to reduce uncertainty about other individuals in order to build relationships, likely accounts for much of this phenomenon.[41] It is not only exceptionally difficult to discern the thoughts and motivations of those who are distally connected to us; as these are our most tenuous connections, we might spend more time analyzing our interactions with them because we are most uncertain about how they think and feel. It's also the reason we feel unease when we are left on read or ghosted in response to our social overtures.[42] Ghosting hits humans at one of our most vulnerable weak spots: our desire to know. We have a *need for closure*.[43] We want to know how things end. We want to understand how the world is working around us. When someone ignores us, it leaves us wondering. And wondering can be maddening. We are left craving answers about the relationship and unable to force them to come. And when those relationships are valuable to us, our anxiety increases as the hours and days tick by.

But the uncertainty related to ghosting is not what really causes humans pain. Certainly we want to know, but more than wanting to know, we crave human connection. Ghosting is a signal of a weak or strained connection. So ghosting actually hits us at an even greater point of vulnerability: our desire to belong and be loved. Ghosting is a big red flag that we might be losing someone we love, or perhaps someone we wanted to love.

This is why people who are ghosted sometimes resort to desperate measures to fill their gaps in uncertainty. They might reach out multiple times to the ghoster, even when continually ignored. They might start to surveil the ghoster on social media. In this case, they are using the internet again as an information source to get bits of information about how the ghoster is moving through the world. The ghoster along with their friends, family members, and workmates form a giant web of connections that is now at one's fingertips. And suddenly the panopticon is flipped. Now the person is in the center, watching the people in the cells around them. The product in the larger sociotechnical machine has now become the guard.

HOW TO SURVIVE THE INTERNET

The Sociotechnical Panopticon
Survival tip #20 We must acknowledge the privacy we lose when we engage in online social networks. Once acknowledged, we either accept the realities of our online worlds and

continue to engage, pull away from participation entirely, or place other limitations on our online engagement. Realistically, however, it's extremely difficult to completely disconnect in a connected world. Unless you go completely off the grid and raise horses in Montana, you're probably going to be tracked.

I <3 My Phone

Survival tip #21   If you think you're addicted to your phone, you probably are. But remember, your phone is the tool that connects you to everything that matters to you in the universe. If you use that tool wisely, it can be a key to unlocking intimacy on all sorts of levels. And if you want to hop on one of those dopamine detox programs that are gaining popularity, just remember that they are meant to increase your response to the activities that give you pleasure, not dampen it.[44]

The Social Support Network

Survival tip #22   Your phone is the handiest tool every invented to help get you the things that you want. Need some food? DoorDash. A ride? Uber. A present for your colleague? Amazon. Almost anything you need is at your fingertips. This is so easy for us to see. Yet we often overlook how much we can use our phones to get the social support we need. It's your bat signal. Your Jedi force. Aside from physical stimulation, like hugs and kisses, your phone is the portal to every sort of social support you'll ever need. Use it wisely.

Tidying Up Your Brain

Survival tip #23   If a four-year-old can learn to control their thoughts, you can too. You choose your thoughts. You control your reactions. No one can *make* you think anything. If you have a negative thought, plan a purposeful counterbalance with a positive one. Worried about an upcoming task? That's fine. Normal, in fact. Just make sure you also spend some time thinking about your next vacation.

Destroy after Reading

Survival tip #24   Ghosting hurts us where we are most vulnerable. Ghosters know this, and yet they still choose to ignore people. If someone ghosts you, they are either playing a game or don't care about you right now. Sure, they might have lost their phone. Of course, they may be super busy. But if someone really wants to talk to you, they will find a way. If they don't, move on. Immediately. Don't stalk them. Don't obsess. Don't waste a single moment scanning the internet searching for answers about whether or not they really care. Life is short, and time is precious.

Bonus Technology Tip

Let's be honest, few of us could imagine an existence without our phones. I could lose my keys, have my wallet stolen, and be dropped alone on a deserted highway, but as long as I have my phone and a signal, I know I'll be OK. It gives me security, connection, and a lifeline to others. This is why I won't ever do a phone detox nor advise anyone else to do the same. It's OK and natural to be dependent on your phone. But when it starts interfering with your life rather than helping you to live a better one, you need to shift. Perhaps decide on some intimate times (e.g., dinner) and intimate spaces (e.g., at night in bed with a partner) when phone use is verboten. And if it's not interfering with your life, don't sweat it. You're fine.

5 HOW TO SURVIVE DATING

THE TALE OF TINDERELLA

I met Ella about twenty years ago. She was singing at a house party in Nottingham, England, shaking a tambourine to an acoustic version of Coldplay's "Yellow," as a roomful of people cheered her on. Petite, with long, straight black hair and an edgy, hippie style, Ella was very much the opposite of me—blond, conservatively dressed, and hopelessly lacking in musical talent. Our meeting occurred through a fortunate circumstance; she was casually dating a friend of my husband's, and she and the other half of her acoustic duo would play impromptu gigs at group gatherings. But over the past couple of decades, our friendship has deepened, and I have been an eager observer of her fascinating life.

Ella has always been a free soul, a beautiful butterfly flitting in and out of adventures and relationships. She's been on a reality TV show, lived on a houseboat in Amsterdam, and toured the world with various bands. Relationship-wise, her life has been similarly adventurous. She's dated primarily younger men, creative types in film and music who share her love of novelty as well as distaste for the mundane. And although she has had quite a few casual relationships, she's also had some serious ones, lasting a few months to a few years, with interludes between these commitments of similar time periods. Many of these relationships were formed via online social networks and dating apps, like Tinder and Bumble, which has led her to laughingly coin herself a veritable "Tinderella."

Like many Generation Xers and baby boomers, Ella is far from a digital native; she got her first computer in college and her first cell phone while in her twenties. Yet the draw of dating applications is appealing to many, regardless of generation. According to a Pew study of US adults, 30 percent report using a dating application, with the highest usage rates among eighteen- to twenty-nine-year-olds (49 percent), then thirty- to forty-nine-year-olds (38 percent), followed by those fifty to sixty-four (19 percent), and finally those sixty-five and over (13 percent).[1] The overall usage statistics for US adults are similar to what we see in studies around the world; a recent review of research across many countries, including the United States, Canada, Thailand, Australia, China, Spain, Belgium, and other European countries, showed that the prevalence rates of online dating application use among adults ranged from 40 to 50 percent for most studies.[2]

The embracing of these technologies among people around the world in almost every age group reflects our almost instantaneous adaptation to new ways of communicating, dating, and falling in love. According to Piaget, adaptation is a specialty of the human condition, reflective of our intellectual ability, and a mechanism by which knowledge is built on across generations. Those who learned how to use these technological tools to effectively find love are thereby advantaged, from an evolutionary perspective, as is our species. Rather than eschewing the current online dating landscape, nondigital natives like Ella embraced it and pivoted. They not only witnessed the evolution of dating but as humans are so good at doing, also adapted to the changed landscape in pursuit of love, sex, and intimacy.

With this adaptation has come benefits and challenges. Of primary benefit is the opportunities this affords people to find someone to love. According to the same 2019 Pew study, 39 percent of the online daters surveyed got married to or had a committed relationship with someone they met on a dating site.[3] Dating sites also appear to be especially appealing to those in particular groups, such as those in sexual minorities (55 percent of whom report using a dating app as compared to 28 percent of those who are straight) and those with college degrees (35 percent of whom use apps as compared with 22 percent for those with high school certifications or less). In short, people have opportunities for love on dating apps, and for specific groups of people, those opportunities are even greater and maybe even more meaningful because of an ability to quickly find people who match them in important ways.

MATCH MADE IN HEAVEN

But is *matching* really a goal of those looking for love? What about the old adage suggesting that "opposites attract"? And isn't matching counter to the architecture of the internet, wherein we are exposed to a wide variety of people from all different backgrounds in all parts of the world?

There are actually two issues here, and I'll address each in turn. Social matching, or *homogamy*, has been studied by economists and those in the social sciences for decades. The dimensions of social status that are typically considered relevant to homogamy are education, race, ethnicity, age, socio-economic status, and religious beliefs.[4] Generally, this research has shown that people who find mates who are like them on these key dimensions, through a process called *assortative mating*, tend to have happier and more enduring relationships.[5] The benefit of matching extends to other psychological characteristics, with those being alike in traits like kindness and cleverness having longer-lasting relationships too.[6]

But matching, of course, seems inconsistent with the structure of the internet, which transcends physical boundaries like geography to allow people more choices with regard to mate selection. No longer are we constrained to date the people in our towns, places of work, or religious or social organizations. We now have the ability to meet and maintain contact with others all over the world, and because of the zero to low cost of online communication, our prospects for deep conversations and emotional intimacy with these others are limitless. The breakdown of these boundaries and moving toward *heterogamy* (people mating with those different from them in these key social dimensions) would, according to researchers from the Institute for Social and Economic Research in the United Kingdom, be beneficial for social mobility, and lead to a more open and democratic society.[7] Instead of the rich and more educated becoming richer and more educated (by like marrying like), increases in heterogamy would result in a more equal distribution of resources across a population. This doesn't seem to be the direction we are headed in, though. Several longitudinal studies in the past few decades have shown that homogamy is still and perhaps increasingly the trend. In a study of data from the United States from 1940 to 2003 examining educational homogamy, or the extent to which couples are similar in their educational attainment, sociologists Christine Schwartz and Robert Mare found that in the most recent years measured, educational homogamy was higher than it had been since 1940.[8] Similarly, data from the British Household Panel Study (using census data from 1971 to 2001 in England and Wales) show that homogamy on most key social indexes is still commonplace.[9]

When mismatching happens, it can have a negative effect on a relationship. One area that has been of particular interest to researchers is a mismatch on educational attainment. In study of 1,083 heterosexual couples in Hong Kong, a mismatch on education seemed especially detrimental to women.[10] Specifically, when wives were better educated than their husbands, it was related to lower levels of marital satisfaction. A US study using Census data from 1980 and American Community Surveys from 2008 to 2012, however, showed that women marrying someone with a lower education than themselves was becoming an increasing trend, due partly to the substantial strides in educational attainment made by women in the past few decades. Despite these advances, women's preference for a man with a higher earning potential than her own remained a consistent trend.[11] Moreover, if a woman has a higher income than a man, they are less likely to marry, and when they do, they are more likely to divorce.[12] This is attributable to many causes, but is most often linked to tensions surrounding nonadherence to traditional gender roles, wherein men are supposed to be the breadwinners of the family. And despite claims that gender equality in some cultures might diminish desires for women to select long-term partners with higher earning capacity (and men to select partners with better looks), a 2019 analysis across thirty-six countries showed that this isn't the case.[13] Across cultures, rather consistently, men prefer mates who are more physically attractive, and women prefer men with a greater capacity to provide resources.[14]

But in the words of US evolutionary psychologists David Buss and Todd Shackelford, "Attractive women want it all."[15] Physically attractive women are not only the most desired in terms of mate value but also have exceptionally high standards when picking their own mates. They want men with good genes (e.g., physically attractive and fit) who have good investment ability (e.g., good earning capacity and income potential), and the potential to be good parents (e.g., want and like children) and good partners (i.e., loving). Attractive women want not only one or a few of these characteristics in their long-term mate; they want all of them. But is the same true for men?

THE CHEERLEADER AND THE ASTRONAUT

One winter when I was in my early thirties, my husband and I lived in the French Alps, in a quaint little ski town called Morzine, which is part of the Portes du Soleil. A six-hundred-kilometer stretch of mountains along the French-Swiss border, the Portes du Soleil contains thirteen ski resorts between Mont Blanc in France and Lake Geneva in Switzerland. Not surprisingly, we skied a lot that winter. And we spent our time off the slopes bundled in a chalet, built into the side of a hill, just a few hundred meters from the village center. It was cozy and idyllic, with hand-painted wooden shutters opening up to the backyard hill where a single goat grazed. Not our goat, I should mention, but like the many things I encountered that winter (e.g., delicious cheeses, weekly farmer's markets, and the Pierre-Chaud stone on which we cooked our meat), I adopted the grazing goat as part of my French Alps identity. Also part of my identity that winter were après-ski meals and nights out in small bars with friends.

On one of these nights out, I posed a question to a few of these men, who were from the United States, college educated, and in their late twenties. I presented to them a scenario that has recently been replicated in actual scientific studies by researchers in China and Belgium. The question I had is summarized nicely by the Belgian researchers in the title of their study: "Are men intimidated by highly educated women?"

As mentioned, previous research related to homogamy and assortative mating suggests that men and women seek others like them on qualities like education and socioeconomic status. Economist David Ong conducted a study in 2015 measuring the profile visits received by 388 fake dating profiles on an online Chinese dating website, which included random combinations of education and income levels for both men's and women's profiles.[16] Notably, neither the women's education nor income level was predictive of a profile click, and yet men with higher educational attainment got more clicks from women. Moreover, women with higher education levels also clicked more often on the profiles of the men with higher incomes. In other words, and in line with what evolutionary theory might predict, men didn't care about a potential partner's income and education, but women did.

More recently, the group of Belgian researchers I mentioned did a similar study, except instead of using profile clicks, they used Tinder and examined connections made to one of 24 fake profiles, again with varied education levels.[17] Using a randomly generated algorithm, they swiped on 150 of the first Tinder profiles that were presented to them, and then they counted how many matches, likes, and conversations each profile received. Similar to Ong's study, they found that women cared about education level and exhibited *hypergamy*, or the preference for profiles with a higher status than their own, whereas men did not. In answer to the question posed in the title of their study, however, they also found no evidence that men were opposed to women with higher education than their own, at least on Tinder.

Now how does this relate to the question I posed to the men in the French Alps? Well, I gave them an almost identical scenario. I told them to imagine they were on a dating website. They see two profiles, each with two photos, one headshot and one full-body shot. It's the same photos in both profiles. Whatever their physical ideal happens to be, this woman is it. In her brief biographical profile, one woman indicates she is a cheerleader for a professional football team, and the other woman notes she is an astronaut. Remember, they contain the same pictures. Same woman. Only the occupation is different. Which one do they click?

Both men in the French Alps said the same thing. I know this is only a sample of two and thus is not at all scientifically valid, but nonetheless, it is a data point. Both chose the cheerleader, not the astronaut. I have great respect for both cheerleaders and astronauts—both do something I cannot do. Nevertheless, as I really thought that interstellar travel was one of the coolest things anyone could ever put in a bio, I was astonished.

There's a man named Jonny Kim who I think everyone should know about.[18] According to his bio on NASA, Kim joined the US Navy out of high school. He then entered the prestigious Navy Seal training program, which is a high-level, special operations position offered to elite candidates. After serving on more than a hundred combat missions in the Middle East, Kim went to college, got a degree in mathematics, and then got a medical degree from Harvard University. He became a doctor. Then, in 2017, Kim applied and was accepted to the Astronaut Candidate training program with NASA. As of an update from 2020, Kim was thirty-six years old and awaiting his first mission. Kim has been successful across so many different arenas—both intellectual and physical; it is amazingly impressive.

It takes a lot of intellectual rigor to be an astronaut. According to the US NASA guidelines, you have to have a master's degree in a science, technology, engineering, and mathematics field or medicine, have completed a pilot school program, or have additional pilot or professional experience. You also must be incredibly physically fit. You have to pass the long-duration spaceflight physical and complete various incredibly challenging physical tasks during astronaut training. Among these tasks is treading water for ten minutes in a flight suit, which in itself sounds like something I could train for years for and still not be able to do. Perhaps because it is both intellectually and physically demanding, being an astronaut is an exclusive career. In April 2021, NASA's website listed only forty-seven individuals eligible for flight assignments.

And yet it wasn't intriguing to the two men I queried.

"Eh . . . the astronaut girl?" they said with a shrug, "too much ambition."

Is this really a thing? After all the suffrage, all the strides, can a successful woman really be seen as having too much ambition? Apparently so, according to some of the men I've queried. Further, as the Belgian researchers noted, "Women . . . shy away from behaviour that may improve their careers in order to avoid signalling undesirable traits on the dating market, such as ambition."[19] So not only is ambition an "undesirable trait" on the dating market but women also are trying to actively subdue their ambition for fear that it might push away potential suitors. Thus while successful men may have to bat away the droves of potential mates, women with similar levels of success may be passed over or even avoided.

I've decided that in this book, I am allowed to have one moment where I can just express what I really feel. Unfiltered. Unscientific. My one raw "share big." This is a book on intimacy, after all. So come close. Lean in. I want to take this moment to say, Are you kidding me? In a world where more than half of women participate in the workforce, are we really not supposed to strive?

But let's get back to Ella and her search for love. Something I failed to mention is that aside from her incredible voice, Ella has a host of other attractive qualities, including a brilliant mind and bewitching beauty. You may have surmised this from what I mentioned already—her reality TV show appearance, and ability to attract and sustain a variety of talented as well as interesting partners—but Ella is an attractive woman. In fact, she's enchanting. Yet unlike the women in Buss and Shackelford's study, she doesn't want the "all" they identified through their surveys. Instead, Ella is looking for a unique blend of personal qualities. In her own words, she wants someone creative, edgy, exciting, active, smart, good looking, and interested in music. Additionally, she states that "it is a bonus if they are spiritual, handy around the house, good at cooking, and great in bed."[20] Ella is successful, and her ambitions, whether she likes it or not, are on public display through the circulation of her music. Also, like the other attractive women who were a focus of Buss and Shackelford's study, she has a long and particular list of traits she is looking for in a potential partner.

All of this is to say that we are not just looking for people but rather *specific* people when we try to find a partner. We are looking for people who match us in important ways as well as those with specific qualities. This makes finding love a challenge, even in the most ideal conditions, like when the world is literally at our fingertips. There are at least three interrelated issues that vex us as we engage in this search online: the "needle in the haystack" challenge, analysis paralysis, and the "plenty of fish in the sea" phenomenon. I will address each in turn.

First, let's go back to the Pew statistic I quoted earlier about 39 percent of people having had a serious relationship or marriage with someone they met on an online dating site. At first glance, this sounds like if you choose to find a mate online, you have a 39 percent chance of finding love. And that sounds like a really great chance. That interpretation is completely wrong, though. And I'll tell you why. But first, let me take you back to high school.

HOMECOMING PRINCESS WINS THE LOTTERY

As cliché as it may sound to some reading this, and as much as it may lend credence to popular US cultural stereotypes, I would have really loved to be the first-year, freshman homecoming princess at my high school. We had a successful football team, and each fall, the homecoming princess candidates (three of them) got escorted in their prom-like dresses onto the field during halftime and were crowned in front of the whole school as part of the big homecoming game. One of these girls was crowned princess, and two of them were awarded the runner-up positions in her court. Although I now see the princess crowning, escorting, and parading through a somewhat different lens, as a fourteen-year-old girl who grew up watching Disney movies and 1980s' romance films, I was enchanted by all of it. Now my class had about two hundred kids in it, about half of whom were girls. This would mean that from a purely statistical sense, I would have had a 4 percent chance of making it into the homecoming princess court. Again, this is wrong. My actual chance was way lower than 4 percent. In fact, it was 0 percent. And the girl who won? Her chance was nearly 100 percent. Why this differential? Because we didn't start out as equals. She had more of the qualities people were looking for in their homecoming princess. She was more attractive, outgoing, and had many friends. Notably, she didn't align herself with any single friend group but instead gracefully navigated her way through friendships across different high school cliques. Everyone liked her, and she was not controversial in any way. She was, by definition, popular.

Online dating, just like getting crowned homecoming princess, isn't like a lottery where everyone has one ball and an equal chance to win the game. It's completely different. Using the lottery example, it goes something like this: a hundred of us enter a big arena where there is going to be a giant lottery to find love. The lottery organizers tell us ahead of time that there will be thirty-nine winners, and you need only a ball to play. Luckily, there are plenty of balls in this arena, and the lottery organizers have a thousand balls to distribute. But they aren't going to distribute them equally. Instead, the number of balls you get is dependent on the extent to which you exhibit certain personal traits. You get some for kindness and understanding, some for physical attractiveness, some for intelligence, some for sense of humor, some for creativity and adaptability, and so on. Now a couple of us will fare well in this lottery and will get a hundred balls, which equals a hundred chances at love. A couple others will get fifty, thirty, or ten. Some will get only a few balls, and some people will get no balls at all. Sitting on the sidelines and watching other people play, it's as if they lost the game before they even entered the arena.

Our chance at finding love is neither fair nor random.

The hard truth is that some people find it quite easy to find love, while others have more of a struggle. Due to the internet's expansiveness, however, both of these groups of people face the needle in the haystack challenge when trying to find love online. About five years ago, I conducted a study I never published, tentatively titled "How Much Love Can Someone Find in One Day?" My friend Anna, who was twenty-five at the time and living near Chicago, populated three popular dating sites with a simple profile (mirroring the basics of Tinder). She had three photos on every site: one headshot and two full length, including one of her holding her dog. Because Tinder has an active "swipe right" process through which one indicates interest in a potential match, Anna planned to use an active engagement strategy on all three sites to level the playing field. On Tinder, she would swipe right, and if matched, immediately send a message stating, "I've looked at your profile, and it would be great to get to know you better." On the other platforms, she would search for profiles she was interested in and send that same message. The plan for the day was simple: between the hours of 10:00 a.m. and midnight (except from 3 to 4 p.m., when we took an hour for lunch) on a single day (we chose Sunday, October 4), Anna would swipe and message everyone whose profile interested her—in this case, men (she's heterosexual) she might consider dating in the Chicago area. Again, to level the playing field, she would spend thirty minutes on a single site and then rotate to the next—a pattern she repeated throughout the day.

During her long day of searching, Anna sent messages to approximately 150 men. She received 601 messages across the three dating sites; 108 men responded to Anna's messages, and 493 men (most of whom she had no interest in dating) reached out to Anna unsolicited. Most of these messages (370) came in the same day, 64 came in the next day, and the rest trickled in over the next few days. Messages varied in length and content from "Hey" to much lengthier and more engaging messages where people would reference her job and interests, and suggest they meet up. Some people even messaged her more than once, including one man who messaged her four times without reply. In her single attempt, Anna received 100 balls for the love lottery. But even with this many balls, it didn't guarantee that any of these men would be a good fit for Anna.

THE PROBLEM WITH TOO MANY OPTIONS

The love lottery windfall might seem like an ideal scenario to some, but when faced with so many options, Anna felt overwhelmed. This aligns with the sentiments of social theorist Barry Schwartz, who has suggested that having an abundance of options—as we do in our technology-filled world—is actually having a negative effect on life satisfaction.[21] A plethora of options makes decision-making difficult and can leave us in a state of *analysis paralysis*, whereby we are unable to move forward because we overthink our options. One of the studies that demonstrates this phenomenon well is the famous jam study by Sheena Iyengar and Mark Leppar.[22] In their field study, shoppers at an upscale grocery store in California were presented with a display with either twenty-four or six types of jam. When there were twenty-four types of jam, more people stopped, but once they got there both groups tasted the same number of jams. Moreover, and here is the important point, those with fewer choices were actually more likely to buy the jam; 30 percent of people with six choices bought the jam as opposed to only 3 percent of those with twenty-four choices.

This study has been replicated with online dating scenarios. Across several studies around the world, researchers have shown that the more options you present someone in terms of dating profiles, the more cognitive effort it requires for the searcher to actually settle on a decision.[23] Furthermore, this can have a negative effect on the quality of one's choice; the cognitive energy devoted to the search actually depletes one's resources for making a good decision, creating a "more-means-worse" effect. But this does not affect everyone in the same way. As with most phenomena, there are individual differences in the ways in which people respond to an abundance of options: *maximizers*, who always want the best, seem to suffer more in the face of options, whereas *satisficers*, who are content with good enough, are not as burdened.[24] In the words of Leo Tolstoy in *Anna Karenina*, "If you look for perfection, you will never be satisfied."[25]

Although initial mate selection processes can sometimes be stymied by the abundance of options, there is an overarching issue that plagues every step of the process: trying to find love, especially online, is like trying to find a needle in a haystack. *Signal detection theory* says that perception is a decision-making process, wherein we need to decipher a signal against a background of noise. Applied broadly to the current scenario, finding love requires you to try to identify the right choice among a sea of foils. The research on online dating I described focused on one aspect of this search: the initial profile like or visit. The search for love is much more complicated, however, and involves a series of decisions, reflecting various stages of relationship formation. At each stage, people must make not only one decision but rather a number of decisions, sometimes about various people. Below is an outline of a dater's potential decisions:

Stage 1: The initial search
> Which venue to find a mate (i.e., which off-line location and/or online application)
> Whether to connect with/message/contact a particular person
> How many people to connect with / message/contact during the same time period

Stage 2: Steps to help determine there is a connection
> Whether to respond to a person
> How quickly and frequently to respond to a person
> Whether to move the conversation to a voice or video conversation
> Whether to meet the person face-to-face
> How many people to chat with or meet during the same time period

Series 3: Determining relationship structure or duration
> Whether the person is suitable for short-term hookup
> Whether the person is suitable for a long-term relationship
> Whether to end a relationship
> How to end a relationship

Each of these steps takes effortful decision-making, which expends valuable cognitive resources that a person might be otherwise devoting to their work, friends, or family. And daters looking for love are not necessarily addressing these steps serially or focusing on a single individual. Instead, they might be deciding whether to message one person based on an initial liking of their profile while at the same time trying to decide if another person (whom they have already met off-line) might be suitable for a long-term relationship. If the sheer volume of decisions is not enough to convince you that finding love is a complicated process, especially in our technology-mediated world, then consider the potential duration of this decision-making process. This process can go on and on, indefinitely, until a dater either decides they want to be alone, or finds a partner (or partners) with whom they connect with so strongly that they decide to give up the search. And at any point in their life, they can change their mind and switch directions.

PSI PHENOMENA AND THE MARSHMALLOW MAN

As if the sheer number of steps were not challenging enough, at each step, there is the potential for error—a chance that you choose something that you don't end up liking. We commonly make mistakes in our appraisals of people that lead us to pursue or entertain the wrong people. Here again, I must provide the caveat that there aren't really any *wrong* people. Instead, what I mean is that in a search for love, the wrong people are those with whom you determine you are not compatible in a relationship sense (either long or short term). But how does this happen? How is it that we make incorrect decisions regarding love?

Two psychologists who have helped me understand why we make mistakes in our decisions regarding others are Daryl Bem and Walter Mischel. Bem was a famed professor at Cornell when I was a student there in the early 1990s. Although I never had a class with him, he was a guest speaker in my introductory psychology class and led the thousand-plus students in attendance through an exercise testing the psi phenomenon (i.e., extrasensory perception). Bem garnered much attention in 2011, when he published a paper in which he offers evidence of precognition (i.e., responses to stimuli before they are shown).[26] This paper sparked intense debate in the field of psychology, spurring calls for it to be retracted, an editorial justification for the paper to be kept in the *Journal of Personality and Social Psychology*, and an onslaught of replication studies in not only the area of psi phenomenon but also the field of psychology overall.[27] In response, Bem and colleagues published a follow-up meta-analysis of ninety experiments in 2015, doubling down on Bem's initial findings and supplying evidence once again that people can anticipate future events.[28] Since then, other prominent researchers have provided empirical evidence in top-tier journals that there is some support for different types of psi phenomena.[29] And yet many scholars are still skeptical, and intense criticism of this work persists. For myself, I've always been impressed by Bem's decision to study psi phenomena. This is partly because it intrigues me to think that human beings might have sensory capacities that are yet undiscovered, but it's mostly because by even studying the psi phenomenon, Bem established himself as a bit of a renegade in the discipline. He goes his own way, despite criticism, which is sometimes difficult for academics to do. And I respected his moxie in doing so.

That said, it isn't Bem's psi research that has really shaped my understanding of human beings. Instead, it is his *self-perception theory*, which posits that people are not great at deciphering their own emotions or attitudes.[30] According to Bem, the internal cues we get about our own emotions are often weak and difficult to interpret. So instead we rely on interpretations of our behaviors to infer how we think and feel. Applied to a dating scenario, when someone asks us how we feel about a new person we are seeing, we might not be able to define our emotions because what we feel is not clear. With no strong internal anchors to guide us, we might say something like, "Well, I went out with him three times last week, so I think it's going well."

Recent research has illuminated, though, that we do have physical reactions that can help us interpret our emotions.[31] A team of scientists in Finland has been working to develop bodily maps of emotions, focusing on the parts of the body that get activated when people feel certain emotions like love, sadness, and surprise. In a 2020 study, which included 6,559 people in 109 countries, researchers found that across cultures, there is much consistency in the parts of the body that get activated when particular emotions are felt.[32] The pattern of activation represents the emotion's "bodily fingerprint," and as we begin to recognize these fingerprints, we are better able to understand our own emotions. Maybe studies into our ability to read other people's minds are too far out, yet the ability to read our own minds seems completely in reach for most humans.

Considered together, these studies suggest that individuals may be able to hone the skill of interpreting their own emotions, but in the absence of obvious bodily cues (e.g., changes in heart rate or respiration), we might have a hard time determining how we feel about someone. Still, even if we can effectively interpret our emotions, feelings can change, and momentary emotions are not necessarily predictive of how we feel over time. Moreover, people are not necessarily the same across different situations, and seeing a person in a different setting might change how we feel about them entirely.

Enter Mischel, creator of the marshmallow test, which measured children's ability to resist the temptation of eating only one marshmallow (or other delicious treat) placed right in front of them for the delayed reward of getting two marshmallows later.[33] Mischel was a strong advocate of the idea that situational contexts were powerful forces in shaping behavior. In fact, he challenged personality theorists of the time who suggested that there was consistency to personality and people could be accurately characterized as having certain types of stable traits.[34] Instead, Mischel argued that people's behavior varies widely across situational contexts. Variations in behavior are not, according to Mischel, random occurrences unreflective of the person's traits. Rather, when people are different in their home than they are in their work environments, or to take a modern spin on the idea, different via text than they are in person, this is to be expected because people can in fact be quite different across situational contexts. Now factor in that text messages are not media rich and text-based communication is asynchronous, and it is easy to see how mistakes can be made when we are trying to figure out if we are compatible with others online. Even if they don't actively try to lie about who they are, others might be different online than they are off-line.

THE ART OF DECISION-MAKING

Regardless of the reason why it happens, making mistakes, even at the beginning of courtship, can be costly. Let's say you choose to message and meet with Suitor #1 based on how they look in their profile photos, or something they say that makes you laugh or turns you on. In meeting with Suitor #1, you don't choose to meet with Suitor #2, who also contacted you, but didn't yet say anything funny or seductive. Unfortunately, the night out with Suitor #1 turns out to be miserable. Maybe they lied about who they were (71 percent of online daters say that people lie online to make themselves seem more desirable).[35] Or maybe, who they were over chat messages was different than who they seemed to be when you met in person. Whatever the case, either they didn't like you, or you didn't like them, and your journey with Suitor #1 is over.

These decisions about what we like and dislike are actually made really quickly. And we need way less information to make these decisions than others would anticipate. University of Chicago researchers Nadav Klein and Ed O'Brien tested this theory using a clever painting judgment experiment.[36] The participants in their study were assigned randomly to one of two groups: experiencers and predictors. The experiencers would see a series of forty paintings, and they would have to decide when, during the series, they had made a decision about whether or not they liked the style of painting. Meanwhile, the predictors had to determine how many paintings it would take for the experiencers to make their decision. After a trial run and seeing thumbnails of the forty paintings (in both groups), the experiment started.

It took only 3.38 paintings, on average, for experiencers to make a decision about whether they liked the style—a number far lower than the 16.29 paintings that the predictors thought it would take for them to make their decision. The researchers then replicated their results across a range of decision-making tasks, and found that people were also quick to make judgments about whether they liked a drink as well as whether someone was a good or bad student, athlete, neighbor, gambler, or even a happy person. Again, in all cases the predictors overestimated how long it would take people to come to their decision. Klein and O'Brien even showed that despite our quick decision-making, we overestimate how much information others need to make a decision about us. In this case, they had MBA students write exactly the number of essays they thought a hiring manager would need to make a decision to hire them. On average, the candidates wrote about four essays, but the hiring managers needed to read only two. We are quick to judge, and others are quick to judge us.

Regarding love, our decisions might be much, much quicker. It takes less than 30 seconds, and perhaps only one-fifth of a second, for the euphoric pleasure areas of our brains (i.e., the same areas activated when one takes cocaine) to light up when we are exposed to images or even just the names of people for whom we have passionate love.[37] Our decision about whether someone is a lifetime partner, however, takes longer. On average, married individuals said it took them 173 days to figure out the person they married was the one. Notably, this is much less time than the 211 days single people thought it would take to make this decision. Still, 30 seconds to feel the euphoria of love and 6 months to figure out if someone could be a lifelong partner seems like a small time investment when considered against the backdrop of a lifetime. So if you feel you make decisions about compatibility rather quickly, you're actually right on track with what science says is normal.

Going back to the two suitors, you've now gathered enough information to determine that Suitor #1 is not *your* one or even one of your ones. So you go back to review your prospects. But when you get back to Suitor #2 (whom you are now evaluating more favorably based on your bad date with Suitor #1), they have already moved on to someone else. And in the unfortunate case that Suitor #2 was actually more compatible with you than Suitor #1, you may have missed a needle in the haystack.

But it's not only actions that can cause us to miss our signal. When we choose to do nothing, we passively allow something to happen, and this decision too can be costly. Consider the classical ethical dilemma, the trolley problem, which finds a bystander who can either do nothing and allow a runaway trolley to hit five people stuck on a railroad track, or flip a switch to make the trolley move to a side track and hit a single person. This dilemma demonstrates well that doing nothing is a decision. Applied to dating, let's say that after the bust date with Suitor #1, you decide you won't break it off immediately. You have your reservations, but you continue to date, chat with, and maybe even enter a committed relationship with them. Choosing to stay in a relationship with someone who doesn't make you completely happy may work out in the end because people can and do compromise as well as change. Yet you might also find that the situation doesn't improve, which could cost you the opportunity to meet someone who will make you happier.

PLENTY OF FISH IN THE SEA

This brings me to a topic I've been examining for almost a decade—one I'll refer to broadly as the plenty of fish in the sea phenomenon. A sentiment expressed by one of the participants in the Pew 2019 online dating study captures the spirit of this issue nicely. According to a twenty-seven-year-old woman in that study, "It's hard to work on a relationship or give a partner another chance when sites/apps are constantly promoting the message that you are surrounded by wonderful singles all the time. It makes you believe that there is always a better or easier option available."[38]

I appreciate this woman's sentiment. A lot. Nevertheless, I think it places too much emphasis on the advertisements of dating applications. You are, in fact constantly surrounded by singles, and I'm certain that many of them are absolutely wonderful. Just how many are there? More than have ever existed in history. In the 2018 US Census, for example, more than 110 million adults reported being single, which was 45.2 percent of all US residents.[39] For comparison, only 28 percent of adults were single in 1960.[40] Meanwhile, worldwide trends over the past thirty years show some of the reasons why single status is at an all-time high: there are more women age forty and up who have never married, more people divorcing, and the age at which people are marrying is going up.[41] From a statistical standpoint, daters have a greater chance of finding someone to love than ever before.

I also think her sentiment places too much stress on the power of external forces for shaping behavior. Is it really that the *websites* are making us believe that there is always a better or easier option, or does our belief in these other options reflect something internal to us?

More than fifty years ago, relationship researchers were examining the reasons why people persist in relationships. The prevailing understanding at the time was based on psychologists Harold Kelley and John Thibaut's *interdependence theory*, which suggested that persistence was based on how much a person depends on or needs their relationship.[42] Whether people choose to stay is based primarily on two factors: *satisfaction* (i.e., How much is my partner fulfilling my most important needs?) and *quality of alternatives* (i.e., How desirable is my best alternative to this relationship?). Later, psychology professor Caryl Rusbult expanded this theory as the *investment model scale*, which is probably the most widely used model of relational investment today.[43] Rusbult included an additional dimension, *investment* (i.e., the resources that are tied to the relationship and would be lost if the relationship ended, like money, friends, and connections), and suggested that investment also influences our dependence on a partner. This dependence, in turn, leads to increased feelings of commitment to stay with our partner long term.

A commonality of both models is that we continually evaluate the quality of relationship alternatives around us. This is normal and commonplace. The fundamental question we face is, Is there, in this world I exist in, a better match for me? Those in committed, satisfied relationships tend to devalue their alternatives, but when satisfaction or investment falter, we might see more quality in the alternatives around us. And when the scale tips, commitment wanes, and this might drive us to actively pursue other options.

For years, Jayson Dibble, Dan Miller, and I have been examining these other options. Who has them, why do we have them, and how much of a threat are they to existing relationships? In a 2015 study, we asked young adults in committed relationships to simply look through their Facebook friends list and count the number of people they would have sex or a relationship with if they were single.[44] On average, young women indicated they would have a relationship with three of their Facebook friends, and they would sleep with about eight of them. Meanwhile, men said they would consider having a committed relationship with eight of their Facebook friends and would sleep with twenty-six of them. And this was only on Facebook. We didn't ask about WhatsApp or Snapchat, the destroy after reading channels through which people might be communicating regularly with relationship alternatives.

We have taken this work a step further as well. In a series of studies, we've asked adults of all ages about whether or not they have *back burners*: people we are romantically and/ or sexually interested in, with whom we communicate with the possibility that we might someday connect romantically and/or sexually. The defining features of back burners are that there is some sort of communication (suggesting a pursuit), and that the admirer has in their mind that they and their back burner may one day have a relationship. In essence, back burners personify the idea of plenty of fish in the sea. Across all age groups, back burners were fairly common. Young adult women reported having four back burners, and young adult men reported eight on these technological channels. I will explore the maintenance of back burners among older, married individuals in the next chapter, but for now I will say that married people have back burners too.

Altogether, these findings were expected. Remember, relationship models indicate that we are continually evaluating our romantic alternatives around us. We were surprised, however, to find that single people did not have significantly more back burners than those in a relationship, and most shockingly, the number of back burners you had, and even the number you communicated with in a sexual way, was not significantly related to your investment in or commitment to your current partner. Now let's think back to the relationship theories I spoke about earlier. Aren't those who are in highly committed relationships supposed to devalue potential alternatives? Wouldn't they be less likely, then, to communicate with those alternatives? To that I say, not necessarily. Due to the current landscape of technological communication, it is increasingly simple to communicate with people to whom we are attracted via various social networking channels. It can take only seconds and be entirely covert. And although we may begin this communication innocently, without any romantic designs, over the course of getting to know someone, we may begin to think, "I could maybe see myself with this person some day in the future." And when that happens, a back burner relationship is born.

Needle in a haystack, analysis paralysis, and plenty of fish in the sea. Technology has changed not only the way people meet and date online but also dating entirely. In a world where people look at hundreds of online reviews to find anything from face cream to vacation rentals, it's difficult to imagine these same people wouldn't use the internet to find and vet someone to love. The internet provides an endless well of information and opportunity—two things that humans crave. And yet at the same time, it can vex us as we try to make those relationships work.

As I'm writing this, Ella has just been through a breakup. They had been together for about 8 months—more than the requisite 173 days they likely needed to make a decision about whether or not they could spend a lifetime together. A guy she thought was *the* guy has told her that he doesn't see their relationship going anywhere. She is heartbroken. So she's doing all the things the internet has told her to do. She has committed to not contacting him or responding to him for 30 days (likely uncomfortable with the uncertainty of ghosting, he's reached out to her five times). She's gone back to her art class and has dropped a new album with one of her bands. She's spending lots of time with friends. With every bold, brave, and beautiful step she takes, she makes a post on social media. "I'm doing great," these posts say. "I'm suffering," she admits to me. Maybe *the guy* will see this all and beg her to come back to him. And if not, as Tinderella well knows, there are plenty of fish in the sea.

HOW TO SURVIVE DATING

The Tale of Tinderella

Survival tip #25   Online dating sites are popular for a reason. They provide opportunity for love. Don't eschew them based on the misconception that they are for the lonely or desperate. They're not. And don't avoid them because you're not of the digital generation. Adapt. I promise that if you do, opportunities will open to you.

The Cheerleader and the Astronaut

Survival tip #26   You might be the sweetest peach, but some people don't like peaches. Don't force a connection with someone who doesn't appreciate your qualities. Are you smart? Find someone who wants smart. Athletic? Pick someone who likes exercise. There are millions of amazing single people in this world. You likely have more suitable matches than you could ever date in your lifetime. Find them. Stop chasing the others.

Homecoming Princess Wins the Lottery

Survival tip #27   Hard fact: not everyone has the same chance at finding love. Want to increase your chances? Dive into the pursuits you love. Deeply. This will give you the best chance to connect with people who share your passions. If you've tried that route and come up fruitless, branch out and develop new passions. Take an art class. Become a spin instructor. Learn to cook. Brush up on your Italian. We can't all be the homecoming princess, but we can all fight like hell to get a chance at the lottery. So fight like hell. Every single ball matters.

The Problem with Too Many Options

Survival tip #28   Recognize how hard it is to date in today's world. Sure, dating was hard when people had to write letters. It was slow, like a pot roast. But today, options go by like sushi on a conveyor belt. This one looks good. Yet what about the next? And what about the pieces that are five minutes around the corner? Just pick a piece of sushi and savor it. If you don't like it, don't pick one that looks like that again. If you do like it, walk away from the table. And stay away from the table until your plate is completely empty.

Psi Phenomena and the Marshmallow Man

Survival tip #29   Try to see new dating partners in as many different environments as possible. Pretend you're on a road trip and aim to visit their top-ten destinations. Text, talk on the phone, and meet in person (in public, with plenty of light and lots of exits). Meet their friends. Go out as a couple. See their home. Have them take you to their favorite restaurant. People might be drastically different as they move through their different spaces. And so might you be. As you go on these dates, get better at recognizing your own feelings. Heart pounding? Butterflies? Maybe a sign that you're sexually aroused. And it might be nothing more than that. What other feelings are important to you? Find them. Trust them. Chase them.

The Art of Decision-making

Survival tip #30   It takes only seconds for our brains to register love. Trust your instincts. If you're not feeling it, don't be afraid to end conversations or dates early. You're not hurting anyone. In fact, you're helping both of you. You're socially economizing and saving your resources (and theirs) for someone who might eventually be your match. You are not a bottomless well of time and energy. Don't float through dating as if you are. Make decisions quickly, trust you've made the right ones, and move on.

Plenty of Fish in the Sea

Survival tip #31   There are two major types of people who fish. One type sits in a boat or on a dock and uses a pole. The other type goes out to the middle of the ocean and uses a really big net. Figure out the kind of fisher you're dating. And if they're holding a net, don't expect to be their only fish. Not comfortable with that? Find another place to swim.

Bonus Technology Tip

The best daters in a tech-fueled era are doing something really important: they are socially economizing. They know what they want, get in quickly, and if it doesn't work, they get out quickly. Time is not an unlimited resource. To make the best use of your time, make a list of your must-haves for a partner, dive in deeply with people you meet, don't make exceptions to your list or ignore red flags, end dates and relationships as soon as you know they aren't working, and don't tolerate bad behavior. Save your energy for the right person. Don't waste it on the wrong ones.

6 HOW TO SURVIVE MARRIAGE

IT'S WHY WE DRINK WINE

I once met a man who told me he could slow down time. He was not a snake oil salesperson or physicist. He was a former US Olympic speed skater who had won the silver medal in the 1994 Winter Olympics. As speed skating races are often won by only hundredths of a second, he had been acutely aware of time his entire life. And he approached me with a question I found irresistible; "Do you want to live forever?" he asked.

"I'm listening," I replied.

It was November 2015, and I had just delivered my TEDx talk in Naperville, Illinois. The man who approached me, John Coyle, was there as a special guest. He had given a talk the year before and was clearly passionate about his topic. He told me to think back to my life as a child and encouraged me think about those long summer days that seemed to stretch on forever. Then he told me to think about the days I'm living now as a wife, mother, and career woman. How quickly did those days go by?

"Quickly. They go by quickly," I admitted.

He then explained to me that those summer days were long and slow because I was having so many new experiences. The beach, friends, the ice cream man, fights with my siblings—I was experiencing things I had never experienced before. Seeing things I had never seen. And this allowed me to think and feel in new ways. And when this happens, when I am in a new world as an eager and emotional explorer, time slows down.

"But aren't I still an eager explorer?" I asked.

"Maybe," he explained. "But your aperture is smaller. You've narrowed your scope and probably made intentional choices to limit your experiences to the things that are familiar. To the things you know you like. The key to stretching time is to continue to expand the depth and breadth of your experiences. Push yourself outside your comfort zone and slow down time."

I watched his TEDx talk again last week. My favorite part was a quote he included by Ernest Hemingway from *The Sun Also Rises*: "Nobody ever lives their life all the way up except bullfighters."

Bullfighting is the absolute antithesis of my life right now. Most weeks, I move in a predictable rhythm, retracing my well-worn path from work, to picking up children from school, to driving the children to sport practices, and returning back home again. Once home, I make chia seed and cinnamon yogurt for the boys, walk the dog, and fluff the pillows on the couch. I am not stretching; nothing is novel. My aperture is very, very small.

A friend of mine reminded me recently that an underlying theme of the movie *Fight Club* is the castration of men, and the rejection of the emasculating grind of marriage and family life that makes men feel like they are dying on the inside. By engaging in underground fighting that no one spoke of, they were regaining their strength and masculinity. They were raging against the confines and cultural restrictions related to modern-day work and family life. They were not fighting against each other. With every punch and cut, they were fighting the doldrum of everyday life.

"And what about the women?" I asked him. "Aren't they also dying on the inside? Women go to work and then spend their weekends driving around in their minivans, taking kids to soccer practice, vacuuming, and making chili for the Sunday football games. None of this is a huge motivator for the sexual dynamism that's expected from wives at the end of each day. Men aren't the only ones being castrated," I insisted, "women are being castrated too. Where is our fight club?"

"Isn't that why moms drink so much wine?" he asked earnestly.

"LIFE IS SHORT. HAVE AN AFFAIR"

On the surface, bullfighting and castration might seem like they have little to do with the topics I covered in my own TEDx talk that day: sex, love, infidelity, communication, and how the internet has changed the rules of the game. But sitting at the intersection between these topics is a state of tension. As highlighted in the last chapter, there is often tension between what we do and what we want. Case in point, even those in committed, monogamous relationships scan their surroundings for available partners. They have back burners.

They communicate with others in romantic and sexual ways. They have affairs. They entertain alternative options. Even though relationship theories like the *investment model* posit that the devaluing of relationship alternatives helps people stay committed to their romantic partners, these alternatives seem anything but devalued.[1] People are still entering the arena, red cape in hand, eager to face the bull.

Time has changed the forum, however. Thirty years ago, the arena was much smaller. Alternative partners met in face-to-face contexts, giving rise to the office romance and other types of backroom affairs. Now, affair partners can be geographically distant, and couples can maintain emotional and sexual affairs for years without even meeting face-to-face. Of course, face-to-face affairs still happen, and from 3 to 30 percent (depending on the study) of those in committed partnerships report having had sex outside their marriage. Moreover, over the past couple of decades, people have gotten more liberal about the acceptability of affairs. From 2000 to 2016, fewer US adults state that this extradyadic sex is always wrong, and more agree that it is wrong only sometimes.[2]

Meanwhile, the rates of online infidelity, including emotional affairs, are climbing. Why? Online dalliances are just so easy. Through my own research with McDaniel, Dibble, and Miller, I have been studying the range of online infidelity behaviors of married individuals, who despite (or because of) their stable, long-lasting connections with partners, venture outside their own bedrooms to make emotionally and sexually intimate connections online. Although websites like ashleymadison.com (sporting the tagline "Life is short. Have an affair") garner much of the negative media attention, common applications and social networking sites, like WhatsApp, Instagram, and Facebook, are enabling millions of people to develop and maintain covert relationships with others online without the hassle of having to send messages behind a paywall. As I've said many times before, you don't need to go to a paid subscription website to have an affair. It will be easier, cheaper, and likely more productive to simply peruse your social media friends list.

Married individuals we have surveyed across the past decade report engaging in a variety infidelity-related behaviors with their "friends" online, including sending sexual messages to alternative partners. *Sexting*—the sending of sexually explicit material via technology—has become a normative practice in today's tech-laden society.[3] It's been coined "the new first base" among young adults, with individuals sometimes exchanging nude photos of themselves before they've even met. The appeal of it is undeniable; sexual images are titillating, and what could be more sexually stimulating than personalized pornography? Despite it being normative and (for the most part) culturally permissible, though, research suggests that sexting is associated with a number of negative individual and relational characteristics, such as ambivalence, conflict in the relationship, and insecure attachment.[4] Most people also consider it cheating; sexting (either sexually explicit or just flirtatious) with another person outside a monogamous relationship rates just below actual physical cheating acts, like taking a shower with someone and intense kissing, in terms of perceived cheating severity.[5]

With whom are our committed partners sexting? Most likely, it's back burners. Those alternative partners I talked about in chapter 5 who are kept simmering on the back burner with the idea that there might be a future romantic connection. But more specifically, who are they? Who are the people who end up on the back burner? In many cases, our biggest relationship threats are people from our partner's past. Recent research has shown that 53 percent of people use social media platforms to check up on former romantic relationship partners, and when asked to identify their most desirable back burner, 44 percent of adults named a past partner, and an additional 24 percent indicated it was a former crush.[6] Whether it is because they have sexual precedence or the *positive affect bias* that occurs over time as we consider our feelings about past relationship partners, the door to past relationships never really closes.

And they walk through that door via the powerful computer they keep in their pocket. Through numerous studies, my coauthors and I have found that those in committed relationships, including married individuals, communicate with alternative partners via common messaging applications like Instagram and WhatsApp. This means that the people with whom our partners are having online relationships are right there in their phone lists, hiding in plain sight. If you've never seen any of this type of communication from your committed partner, it doesn't necessarily mean they aren't doing it. Online cheaters are often clandestine about this communication. Most report hiding messages from their partner.[7]

But these online escapades sometimes have huge costs. In their search for outside connections and stimulation (including pornography), married and partnered individuals are frequently missing out on the intimacy-building opportunities that exist in their own homes. They are choosing the short-lived glimmer of Facebook fantasy to the touch and intimacy of their own committed partners. Perhaps because of this, online affairs can also sometimes lead to relationship dissolution. In the words of divorce attorney James Sexton, anyone who uses Facebook is "an idiot," and by engaging with social media, "It [Facebook] will f-ck with your head, your heart, and your relationship."[8] Media reports support Sexton's warnings; social media like Facebook and WhatsApp have been cited as contributing to anywhere from one-fifth to one-third of divorces, depending on the source.

As I discuss how to survive marriage over the next few pages, I will be careful to avoid repeating the themes I covered in the last chapter. But know that some of the same challenges people face in the world of dating are confronted by couples trying to hold it all together. Married people understand there are many alternative options, they are constantly surrounded by those options, and often find it hard to leave the table as the sushi goes around on the conveyor belt. Perhaps because they are entrenched in an intimacy famine. Or maybe it's because they know that when they walk the straight and narrow, they deprive themselves of the ecstasy of fighting the bull.

AN UNDERACHIEVING UNION

In order to survive marriage, you must first be married. But the likelihood of you being married is the lowest it's been for decades. The United Nations maintains World Marriage Data statistics for 232 countries and areas worldwide—a database that shows the marital status of members of the population,

delineated by age and sex.[9] This comprehensive set of statistics contains data from 1950 to the present day, and more
countries are being added as the survey expands. In its 2016
summary, which included marriage data from 1970 to 2014
across 231 countries or areas of the world, the United Nations
noted that over the last few decades, the number of married
people has been declining.[10] It attributes this to several trends,
some of which I touched on briefly when I mentioned the
rising number of singles in our population.

First, people are delaying marriage. From 2000 to 2014,
the average age of first marriage was 23.4 years for women
and 26.5 years for men, which is two years later than the
1970–1999 statistics (21.3 years for women and 24.3 years
for men). This trend occurred across every major region of
the world. Second, cohabitation and other types of unmarried
unions are becoming more common. Although this was once
a trend mainly among young adults, cohabitation is becoming more popular among all age groups, especially in certain
areas of the world, such as Latin America and the Caribbean.
Third, across almost all age groups, fewer people have ever
been married than in past decades. Notably, more than 90
percent of people worldwide will have married at least once
by the age of fifty, but the number that have ever been married has declined, especially among those under the age of
thirty. In the twenty-five- to twenty-nine-year-old age group,
for example, the percentage of those ever married has declined
10 percent for both men (from 70 to 60 percent) and women
(from 90 to 80 percent). Finally, there are more people who
are divorced, separated, or widowed than in previous decades.
This trend is attributable mainly to women, though; men

have shown no such increases, likely because women tend to marry at younger ages and get remarried more often.

Although low rates of marriage are not inherently problematic, an associated issue, *unrealized fertility* (i.e., unrealized aspirations of fertility), has become a major concern in recent decades as the number of children that people are having in many industrialized nations is falling below replacement levels.[11] Unrealized fertility does not mean simply not having children; instead, it reflects a disconnect between what a person wants and what actually happens. It is also sometimes framed as "underachievement," which is a less sensitive term reflecting essentially the same thing: a person has fewer children than they wanted, and this might include having no children at all. A survey of women at the end of their reproductive years (forty-four to forty-eight years of age) across seventy-eight countries showed that unrealized fertility was fairly common.[12] Depending on the country, between 20 and 50 percent of women reported having had fewer children than they desired, and 10 to 20 percent still wanted another child. These rates were highest in sub-Saharan Africa and lowest in South Asia.

But how many children do people want? Around the world, the ideal number varies, but the most desired family structure in Europe is still the two-child one.[13] This is also true in the United States. As shown in an analysis of data from the 1979 cohort of the US National Longitudinal Survey of Youth, most of the people surveyed in their teenage years indicated they wanted at least two or more children (88.2 percent of men and 82.7 percent of women). People in general desire children. This aligns well with recent trends from the Pew

Research Center: rates of women in their forties who have had children dipped from 90 percent in 1976 to 80 percent in 2006, but rebounded to 86 percent in 2016.[14] Overall, people want to have children, and they do.

That said, unrealized fertility still occurs. It may happen for a few reasons, including fertility problems (which affect 16 percent of married couples in the United States, and one in four couples in developing countries), competing obligations (e.g., career, leisure, or education), and the social and biological complications that sometimes occur when individuals choose to have children at older ages.[15] In pursuit of other goals, some women simply run out of time. Delayed marriage has been cited as one of the reasons for this, but shifts in women's engagement in education and the workforce have also been identified as contributing factors.

Marriage and having children are not necessarily essential for intimacy and a happy life. In fact, having children can bring about significant struggles among couples as they learn to navigate multiple roles (e.g., wife and mother, or husband and father), try to maintain emotional and physical intimacy, and make compromises about how to parent their children. The question of whether the net effect of children is positive has been studied for decades. According to 2016 research, whether or not children bring happiness is at least somewhat dependent on the country in which you live.[16] In countries like Portugal and Sweden that provide extensive family support, such as paid parental leave and subsidies for childcare, parents are generally happier than nonparents. In other countries, however, like Greece, the United Kingdom, and United

States, parents report being less happy than nonparents. In fact, of twenty-two industrialized nations, the United States had the greatest negative discrepancy between parents and nonparents in terms of happiness. Simply stated, US parents are the unhappiest as compared to their childless counterparts.

Meanwhile, research on marriage frequently points to its many benefits, especially for men. As mentioned in previous chapters, being married seems to be a protective factor for loneliness and isolation. It gives you resiliency to illness and early death. This may be due to the fact that healthier people are more likely to get married, but it might also be due in part to the beneficial social activities associated with marriage, including partners encouraging each other toward healthy eating or exercise.[17] It provides a convenient companion too—a person with whom you can spend your time, grow, and have experiences. Marriage provides you with a witness to your life. People who are married tend to be happier, healthier, and economically better off than those who are unmarried.[18] Yet not all marriages are happy, and 50 percent eventually end in divorce. Moreover, even for those who stay married, happiness does not seem to last.

THE FIRE IS THE FLOOR, NOT THE CEILING

At one point, one of the most dependable trends in social sciences was the U-shaped curve of marital happiness showing that it declined after marriage and then slowly rose as people neared retirement. But that was based on cross-sectional data, and more sophisticated analyses are not as optimistic. In fact, marital satisfaction seems to decline through the entire course of a marriage.[19] It sharply decreases in the first few years, and depending on the model, either continues to decline over the next fifty years or stabilizes after about twenty years of marriage to a level significantly lower than what it was at the start. The reason why? Disenchantment.

Marriage is hard. Two different people try to combine lives, and there is an adjustment period as they attempt to navigate the challenges of the role shift and conflicts that often arise during long-term relationships. At the same time, individuals in couples may not always grow together. Instead, they might grow in different ways that make it even more difficult for them to agree as well as find common bonds and interests. Keeping in mind that *social exchange theory* dictates that the decision to form and sustain relationships is dependent on the benefits and costs that one perceives in the relationship, these changes might shift the balance out of favor for one or both of those in a romantic couple.[20] Finally, just like hugs may give you fewer butterflies over time, the effects of habituation might affect marriage. Over time, people may simply take each other for granted, and the behaviors that provided excitement and stimulation at the beginning of the relationship may start to lose their luster.

There are many theories of love, but the one that I and many other researchers reference most was developed by esteemed US psychologist Robert Sternberg, currently a professor of human development at Cornell University. Sternberg's triangular conceptualization of love delineates different components of love, depicted metaphorically as three points of a triangle.

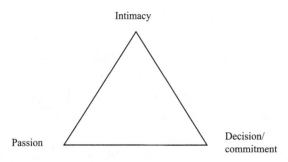

Intimacy

Passion

Decision/
commitment

According to Sternberg, *passion* is the heat, fire, and butterflies—the drives that lead to attraction and arousal in the relationship. *Intimacy* is the closeness, sharing, and feelings that make one feel warmth toward the partner. And *decision/commitment* reflects two different components based on the decision to love or couple up with someone, and the commitment to stay with that person over time. As Sternberg notes, the two components of decision and commitment are not necessarily synced, as people sometimes make the decision to couple up without desiring to commit in the long term, and there are also people who are committed who are not in love.[21] An important feature of this theory, one espoused by other theorists as well, is that there is a delineation between the intensity of passion and the warmth of companionate or intimate love.[22] But the ideal love, in Sternberg's view, is consummate love, where an individual enjoys all of these components of love in their relationship.

How frequently do people have this fulfilling and complete consummate love? And how does it relate to couple satisfaction? A balanced weight of all three components might be hard to achieve. Sternberg suggests passion peaks quickly at the beginning of the relationship but fades over time. In fact, it usually drops off quickly.[23] In contrast, when people choose to stay in relationships, commitment tends to grow. Intimacy, meanwhile, has a nonlinear pattern; it grows at first, though more slowly than passion, and then decreases over the course of a relationship. Although research has been inconsistent in supporting the pattern of how these three components of love change over time, a 2020 study involving more than seven thousand participants across twenty-five countries showed that passion was strongest in couples with the shortest relationship duration, and commitment was strongest among those together longer.[24] Additionally, among couples of all durations from under a year to twenty-one years or more, people reported having much less passion in their relationship than either intimacy or commitment. Passion not only wanes but also seems to suffer throughout the course of a relationship, except perhaps at the beginning.

Yet passion is important in romantic relationships. It is related to couple satisfaction and is a quality people seek in a potential romantic relationship.[25] Passion can also promote a positive and healthy sex life. Considering this, you may find it unsurprising that the frequency of sex appears to decline over the course of a relationship too. In one recent study, seventy-two newlywed couples from the midwestern United States were followed for a period of four years.[26] At first they reported having sex about nine times a month, but at the end of four years, husbands and wives reported having sex less than six times per month. This equates to a more than 33 percent reduction in sex in only four short years. And these were newlyweds! Sex does not decline equally in all couples, however.

A 2019 study involving ninety-two young adult couples in Norway showed that when wives have a more open *sociosexual orientation* (i.e., attitudes toward casual sex and sexual desire), sex lives tend to stay more active.[27] It was wives' and not husbands' sociosexual orientation that mattered because in heterosexual couples, women are the ones who more often determine the frequency of sex.[28]

What happens over the course of ten years? Or the next fifty? Unfortunately, the frequency of sex also declines steeply in midlife and later. As shown in table 2 below, a study using two national data sets in the United States involving individuals age forty-four to seventy-two years of age demonstrated that the frequency of sex decreases over one's lifetime.[29] In old age, those fifty-seven and above reported having sex less than one time per week. In the samples analyzed, most of the individuals were married, but the frequency of sex was lowest among those who were unmarried, many of whom did not have sex at all during their senior years.

Passion and sex declines. Commitment tends to grow, however. Yet many relationships end. When they end, I am

Table 2

	Men		Women	
	44–59 years	57–72 years	44–59 years	57–72 years
Sexual frequency (times per month)	6.18	3.13	4.68	1.74
Sexually active %	87.8	72.0	71.9	45.5

Note: Data retrieved from Amelia Karraker, John DeLamater, and Christine R. Schwartz, "Sexual Frequency Decline from Midlife to Later Life," *Journals of Gerontology: Series B* 66B, no. 4 (July 2011): 502–512, doi:10.1093/geronb/gbr058.

often surprised at a common sentiment I hear people express: "I love him, but I'm not *in love* with him." In this case, when they say they love their partner, I'm guessing this means they feel their relationship has high levels of intimacy. But what do they mean by "in love?" I often wonder whether their conceptualization of being "in love" is really just the intense heat of passion. Perhaps when people say they are no longer in love, they mean that the sexual attraction, lust, or infatuation they felt towards their partner at the beginning of the relationship is gone.

Those butterflies people feel, of course they fade. Early in a relationship, those feelings might be indicative of trepidation. Research has shown that people sometimes misattribute their feelings of anxiety as feelings of arousal. In fact, there's a famous study by Canadian psychologists Donald Dutton and Arthur Aron that illustrates this phenomenon well. In their experiment, some men were approached by a woman on a high, shaky bridge and other men were approached on a low, stable bridge.[30] Men on the shaky bridge were significantly more likely to put sexual content into a story and call the woman than those on the low, stable bridge. When the woman was approached by a male, though, there were no significant differences between the two groups of bridge crossers. According to the authors, the adrenaline sparked by the shaky bridge crossing translated into misattributed feelings of sexual arousal. This might be why dates on shows like *The Bachelor* (a television program in the United States that now exists in some form in a number of countries worldwide) sometimes involve extreme heights and scary activities. Understanding that anxiety might be misconstrued as arousal, producers are actually priming their contestants for the midair bungee jump make-out sessions.

And what about the intensity of first kisses, the shudder of your body when it's being touched by someone new and attractive? These reactions might be attributable to novelty, exciting the previously cited dopamine-laden reward pathway. It feels intensely pleasurable. Euphoric. And those feelings can lead to giddiness and infatuation. But over time, as you habituate to your partner and their touch, the euphoria fades, slowly being replaced with the stability of commitment. Perhaps this is why there is a marked decrease in sexual frequency as well as satisfaction in couples over time.[31] Commitment replaces passion. One way to maintain interest and passion in a relationship may be to recognize when these internal shifts in excitement and intense pleasure occur. Then, instead of thinking "I've fallen out of love," people can make behavioral changes in their relationship to correct course. Remember—sexual activity is a behavior not a feeling. But when people lose their feelings of attraction and sexual interest in their partner and stop having sex entirely, it can often spell the end of a relationship. And this is connected to a larger issue: many couples are not even going to bed together.

TAKE ME TO BED OR LOSE ME FOREVER

Bedtime is an important time for most couples. It's a time to relax, unwind, and share leisure together, largely unencumbered by the other demands of life, like children, work, and house commitments. In fact, it might be the only time of day when couples can truly be alone. Perhaps because of this, during bedtime routines, couples frequently share intimate moments, both physical and emotional, and the emotionally intimate conversations they have can help them strengthen their identity as a couple.[32] Yet what happens when people don't share these emotionally and physically intimate moments? When their bedtime routines are out of sync or undesirable? Might this translate into *unrealized intimacy*? And how does this relate to couple satisfaction?

This is a question my coauthor, McDaniel, and I set out to answer in our latest research.[33] We asked 289 people living in the United States who were in long-term cohabiting relationships (average 9.67 years) to describe their typical bedtime routines. We encouraged them to be as detailed as possible in describing where they were, and what activities they engaged in with and/or without their partner. Individuals gave wildly different responses. Some expressed that during their typical routines, they spent time together, engaging in intimate activities:

> Our usual ritual is that we both brush our teeth but not always at the same time. We lay in bed and goof off a bit on our phones while maing [*sic*] small talk and then we hold each other for a bit and talk about random things until both of us are ready to turn off the light and sleep. Then we snuggle and roll over until we sleep. [P1]

But many others indicated they spent little or no time with their partner during bedtime:

> We are in the same house, he watches sports and then a movie, sometimes we'll watch a movie together, then I fall asleep on the sofa and he goes to the bedroom and falls asleep after he's done surfing the internet. [P2]

She's in bed way before me and is asleep by the time I'm in bed. [P3]

She would mainly take a shower, watch tv and then go straight to bed. No kissing no sex. [P4]

So that we could measure unrealized intimacy at bedtime, we then asked couples to indicate what they would like to do in their ideal bedtime routine, again giving them the instructions to be as detailed as possible. Here their answers were very different than what they described in their typical routines, as illustrated by the responses provided by the same four participants described above:

We would be together of course and we would enjoy some down time to relax. We would then go to bed at the same time after brushing our teeth together and settling other night time engagements. We would sit down in bed and pray together and we would hold hnds [*sic*]. Then we would smile at each other and enjoy some quiet time together just relaxing. After that we would hold each other tightly and talk about our goals and dreams in life and just enjoy the moment together. Then we would go to sleep. [P1]

To get in the bed and listen to music to fall asleep, cuddle and then chit chat a bit, then fall asleep. [P2]

Be in bed together by 11:30, talk for a bit, maybe read a little. Snuggle for a bit, I'm touchy-feely guy. Then pass out. [P3]

We would talk, watch a movie. Probably drink wine and then make love. I would like to have it with us performing oral sex on each other. [P4]

The contrasts are striking. Many individuals described a large disconnect between what they wanted and what they actually did during their nighttime routine. As an example, slightly more than half (59 percent) indicated in their narratives that they typically went to bed with their partners. And in terms of unrealized intimacy, 27 percent indicated that would want to go to bed with their partner in an ideal bedtime routine, but this wasn't typical. But they were missing out on physical and emotional intimacy too. More than one in four (27 percent) indicated they wanted physical intimacy (from hand-holding to sex), which was not currently a part of their typical routines, and one in five (20 percent) indicated they wanted some type of emotional intimacy, also not currently a part of their bedtime routine. If we extrapolate these figures to the entire US population (more than sixty-two million couples in 2020), that means that millions of couples are going to bed each night alone, without the physical and emotional intimacy they crave.

Of course, the most important question we wanted to answer was whether these bedtime routines related at all to the participants' satisfaction with their relationship or life in general. They did. Simply going to bed with a romantic partner predicted bedtime satisfaction, as did having emotionally intimate moments, such as talking about their day and engaging in joint technology use like watching TV together. In turn, increased bedtime satisfaction led to more sexual, relationship, and life satisfaction. Regarding unrealized intimacy, when there was a mismatch between their ideal and typical bedtime routines in terms of physical intimacy, it had a negative effect on bedtime satisfaction.

Now let's consider these findings in light of Loewenstein's "more sex" study, which I mentioned in the introductory chapter.[34] Recall that in that study, happily partnered individuals who were asked to have more sex did not experience increases in happiness. In fact, the opposite happened. It made their moods worse, and their desire for their partners waned in comparison to a control group. Again, perhaps the researchers targeted the wrong sample. Maybe we need to find people who are unhappy in their current relationships. More specifically, we need to find people who are unhappy because of a sexual desire discrepancy.[35]

I WANT YOUR SEX(UAL DESIRE)

At the level of the individual, *sexual desire discrepancy* occurs when the frequency or intensity of sex one is having does not align with the sex they want. Within a couple, this happens when one partner wants sex more often or at a different level than the other. Sexual desire discrepancy among couples has become such a major issue that in 2020, the European Society for Sexual Medicine, composed of a team of psychologists from Portugal, the Netherlands, Italy, and Croatia, released a position statement on the topic.[36] In doing so, the team wanted to address some of the limitations on the topic, both clinically and empirically, and provide an expert statement to help guide future interventions.

One of its major positions was that *sexual desire discrepancies should focus on the couple and not the individual.* Although sexual desire has biological underpinnings in terms of hormonal and neurotransmitter activity, the initiation of sex and responses to sexual overtures from a partner have little to do with biology, and much to do with couple dynamics.[37] Sexual responsiveness might also be rooted in larger societal constructs like religious beliefs and sociocultural factors. According to the European Society for Sexual Medicine, sexual desire is often conceptualized as an individual trait, with people being categorized as having high and low levels of sexual desire. But this is problematic on many levels. First, if a person is categorized as having low sexual desire, they are often pathologized, especially within the context of romantic relationships, where sexual responsiveness is expected. Second, in keeping with the team's position that sexual desire discrepancy is a dyadic issue, categorizing one person in the couple as having a low sex drive ignores the gamut of interactions that might be affecting sexual responsiveness. Finally, and this is a point largely unaddressed by the European Society for Sexual Medicine, it is women's sexual dysfunction and low sex drive that are frequently highlighted as culprits for declines in couples' sexual activity.

In heterosexual relationships, women are often considered the gatekeepers of sex, but this is not exactly true. It is true that men think about initiating and actually do initiate sex more frequently, yet when these initiations are controlled for, there are actually no differences between men and women in how often they refuse sex.[38] So *gatekeeper* is most certainly a misnomer. That said, it is still the case that men tend to want and initiate more sex in a relationship. In a small number of cases, this might be due to pain or discomfort. A large minority of premenopausal women (21 to 28 percent) report having sexual dysfunction, related to issues like lubrication, orgasm, and pain.[39] But this also means that approximately three-fourths of premenopausal women report no sexual dysfunction. So why the large gender differences in desire? Numerous studies have shown that men typically have a stronger sex drive than women; they think about and want to have sex more often, and have more intense sexual fantasies.[40] Men are also much less likely to report having a low sex drive that causes them distress; approximately 30 percent of women have hypoactive sexual drive disorder as compared to only 15 percent of men.[41] Attributable at least somewhat to greater levels of testosterone, men are more likely to exhibit strong levels of sexual desire than women. Within couples, this trend is evident too.

Consider the findings of a 2019 study by US psychologist James McNulty and colleagues. In two studies, they followed newly married adult couples in Ohio and Tennessee for approximately four years. Their goal was to determine what happened over time to sexual desire, as measured by questions like "I enjoy thinking about having sex with my partner," and how that related to marital satisfaction. Initially, both husbands and wives rated their sexual desire at levels above the scale midpoint. Yet aligning with the finding that men typically have greater levels of sexual desire than women, even at the beginning of the relationship, husbands rated their sexual desire for their partner higher than did their wives. Still, this wasn't the researchers' most notable finding. Husbands' sexual desire remained relatively high and stable over the course of one year and then four years. Wives, on the other hand, noted that their sexual desire for their husbands diminished significantly and rapidly, with declines happening even during the first year of marriage. Simply stated, men continued to want their wives in a sexual way, but over time, women wanted their husbands less. More important, when sexual desire waned for wives, it had a negative effect on relationship satisfaction for both partners.

These researchers wanted to know why sexual desire declined among women. One factor they examined was childbirth. Many of the couples had children over the course of the study, and childbirth did contribute to decreases in sexual desire among wives. Perhaps, the researchers posited, having a child adds additional life stress, and this stress might serve to dampen sexual desire. Any person who has gone to bed at night exhausted, shirt soaked with food stains after tending to a child all day long, would agree that this indeed might be part of the problem. But why wouldn't this affect both husbands and wives equally? Aren't they both parents? According to the researchers in this study, this stress is particularly pronounced for women because they typically bear more of the burden of child-rearing than do men. Alternatively, the researchers suggested that differences between husbands and wives in decline in sexual desire might relate to gender differences in reproductive strategies. Women, who have a larger parental investment than men, including the period of gestation, the actual birth, and feeding, might be less focused on future reproduction once they are pregnant. Men, meanwhile, with a lower level of parental investment from an evolutionary perspective, might seek to maintain their sexual activity to increase their fitness.

Parenting isn't the entire answer, however. Even in couples who didn't have children, wives reported declines in their sexual desire for their husbands. According to the researchers, this might be related to mate attraction strategies. Apparently, sex serves many functions, including promoting pair bonding.[42] When women have sex outside their fertile window (i.e., extended sexuality), it might be due to desires for pair bonding. This idea is supported by research that shows that when women perceive low investment from a romantic partner, they are more likely to engage in this extended sexuality.[43] Conversely, when wives know that their partner's investment is secure, as might be evidenced, for example, by time spent in a marriage, their sexual desire declines. In an unfortunate twist of fate, once a wife secures a commitment and her husband becomes more invested, it may dampen her passion for him. Time, it seems, is an enemy of women's desire.

HOW THROW BLANKETS AFFECT SEXUAL DESIRE

But it is not actually time that is the enemy of desire; it's familiarity. Think back to the beginning of your current or most recent romantic relationship. What did sex feel like at first? Go ahead. Take one, long, luxurious moment to think of that hot, fresh, start-of-the-relationship sex. Now think about your current sex life. The mechanics of sex may not have changed much. In fact, the mechanics may have improved, as you and your romantic partner learned more about each other's bodies, and became more comfortable expressing your sexuality and embarking on sexual adventures. And yet you may not respond to your partner's sexual touch the way you did when the relationship was new. What has changed?

Well, you've likely been touched by that partner hundreds, if not thousands, of times. Just like habituation diminishes our response to other types of stimuli, like a blinking light or violence on television, habituation may make us less responsive to sexual stimuli too. Habituation might also account for our desire for novel partners. The *Coolidge effect*, or the renewed sexual interest in a novel partner after sex with the still-available partner(s), may be due at least in part to habituation. This effect, which has been studied mostly in males but has been shown to relate to female sexuality as well, has been attributed to the effects of novelty on the dopamine-rich mesolimbic pathway.[44] Novelty sparks desire and intensely good feelings. But once that novelty disappears and someone habituates to their sexual partner, it takes more of the stimulus to get the same reaction. Or it takes a new partner entirely.

I know a twenty-something woman, Ashley, who didn't want sex with her former partner when the throw covers on the couch were not folded properly over the armrests. She expressed to me that it turned her off completely. To those who care nothing about carefully folded throws, perhaps this seems odd or nitpicky. But the throws were just a symbol. According to Ashley, she asked for little in their relationship in terms of her partner's contributions to the housekeeping tasks. The one thing she asked of her partner was, "When I come home at the end of the day, please make sure the throws are folded." Blankets left strewn about the couch were a symbol of his disregard for her feelings, which darkened her mood and diminished her libido. Lust and attraction are not solely about biology; they can be turned on or off by external factors.

Note that I said Ashley's former partner. They were together ten years, but she broke off their engagement three years ago. One of the main causes? A dysfunctional sex life. But more than that, as evidenced by the throw blanket example, it was dysfunctional communication. In this way, familiarity is the enemy of attraction too. Attraction and lust are two of the three emotions proposed to be characteristic of mating (the third being attachment, or the desire for close contact).[45] Over time, as people have more experiences together, attraction might grow or fade. This may be influenced by a variety of factors, including how physically attractive we perceive them to be as well as dynamics within the relationship.

Sexual attraction does not occur in a vacuum. Attraction is fueled and quelled by other interactions throughout life. Consider, for example, how much you're attracted to your partner when you're in the middle of an argument, or when you're stressed and focused on an issue with work or child-rearing. Psychological distractions can make us turn away from romantic and sexual pursuits, and turn toward the other issues that occupy our headspace. And some of these distractions—like the throw blankets on the couch—might seem to be fairly trivial. To Ashley, the issue was not only critical but also symbolic of bigger and deeper problems in the relationship. I'm happy to report that Ashley is now in a new relationship. And she's happy, their sex life is phenomenal, and the throw blankets are always nicely folded.

There is no secret recipe for a happy marriage; the secret to surviving marriage really comes down to compatibility on core values and interests along with effective communication. When these aren't aligned and communication falters, counseling might help. But if a person has turned off and disconnected, reviving the relationship might be nearly impossible.

In this case, rather than try to fix the issue, some couples might decide to call it quits and start over with someone else.

And just like that, we've come full circle. When people are unhappy in relationships, they usually have the option to leave. But sometimes they don't leave and instead choose to fill their voids with other things. Like back burners. Or affairs. Alternatively (or additionally), they turn to pornography to fill their needs. In response to this demand, the online sex industry is getting more and more tailored as well as sophisticated. Although it cannot be said that sex is the sole force that drives the technology industry forward, conservative estimates suggest that approximately 10 percent of internet traffic is pornography.[46] Moreover, an army of AI-fueled, talking sex robots, with lifelike bodies, skin, and human body temperatures, are pushing the entire field of robotics forward in terms of sophisticated animatronics.[47] Looking ahead, the adult industry boasts ambitious goals, with fully immersive virtual reality sex, connected sex toys, and sex robots on the horizon.[48] If an army of robots is set to take over the world, they might conquer the world of sex first.

Even for the layperson who won't ever own a sex robot, technology is changing the way we have sex. Pornography is shaping our cultural views of sexuality. It is an outlet for sexual stimulation. It provides the novelty people crave devoid of the commitment and investment that can sometimes sap desire. Although the consumption of virtual sex can create dissonance and guilt, and sometimes conflict within couples, some people perceive pornography as the only way they can meet their sexual desires.

I suppose I'm not the only one who is moving through life in a familiar rhythm. There are millions of people marching through the world to the consistent beat of marriage and family life. For some of them, it feels glorious. Commitment provides a strong foundation for the ultimate, consummate love. For others, it feels like mundanity, a small aperture, life lived quickly without much novelty. And for others still, for those whose beat is drowned by the clanking of desire discrepancies, unrealized intimacy, and deep unhappiness within their romantic relationships, life can feel much, much heavier. For all of them, perhaps a bit of novelty in some form will make them feel, for just a little bit, that they are once again in the arena, fighting the bull.

HOW TO SURVIVE MARRIAGE

It's Why We Drink Wine

Survival tip #32   Marriage and family life are not necessarily full of novel experiences. You might move in familiar rhythms, which might feel comforting to some and depressing to others. It's crucial to figure out where you lie in this spectrum, and perhaps more important, where your partner lies. We don't all have to live our lives like bullfighters, but it's nice to have some moments of novelty and excitement interspersed between the soccer games. You won't stumble on them. You'll have to create them. And if you create them with (rather than apart from) your partner, you're much less likely to find them in a basement somewhere in a secret boxing match or downing a bottle of wine.

"Life is Short. Have an Affair"

Survival tip #33 Infidelity can take many forms. It might be actual sexual intercourse, but it could also be sexting or forming an intimate connection with an attractive person outside your primary relationship. These online dalliances are often easier to find, maintain, and cover up than actual off-line affairs, which might explain their rising prevalence. Although most people think that emotional intimacy and sexting are cheating too, not everyone does. So make sure you talk early and often with your partner about what they think constitutes cheating. If you don't align on your beliefs, negotiate, or find a partner who's on your same page.

An Underachieving Union

Survival tip #34 It would probably take you only one day in the shoes of a childless, unmarried adult for you to feel how much emphasis most cultures place on these milestones. People are usually pressured, strongly, by parents and larger sociocultural forces to get married and have children. Yet many people are delaying both or even eschewing them entirely. Although marriage does appear to provide benefits, especially for men, having children and getting married also bring significant amounts of stress for a couple. You don't have to be married or have children to have a happy life. Ignore the propaganda and do what feels right for you.

Fire Is the Floor, Not the Ceiling

Survival tip #35 A subtle shift in our concept of love might make a big difference in marital happiness. More specifically, people need to start seeing the fire as the floor, not the ceiling. Instead of idealizing the early, first steps of young love, let's start celebrating the milestones of long-term relationships à la "I'm so happy I no longer feel nervous and full of butterflies. This must mean I really trust and am comfortable with my partner." Once you get there, you've won. Stop thinking that you've lost something. All you lost was fire, and fire burns.

Take Me to Bed or Lose Me Forever

Survival tip #36 If you take only one piece of advice from this book, let it be this: go to bed with your partner. This might mean a huge sacrifice on one or both of your parts. Maybe this means the person who usually goes to bed sooner has to endure the TV or a laptop being on for a while, or the person who goes to bed later has to turn in just a few hours after they get home from work. It will be worth it, though. This simple fix alone might increase your happiness as a couple. Being in bed together might also prompt other bonding activities like talking or having sex. And if you don't want to talk to or have sex with your partner, then you should probably go to counseling. Or find a new partner.

I Want Your Sex(ual Desire)

Survival tip #37 Lots of couples are disconnected in terms of what they want and what they actually get in regard to sexual activity. This is attributable mainly to women, whose sexual desire for their partners appears to wane over time due to a variety of factors, including child-rearing. Yet people can engage in sex without sexual desire. They can even do so in the absence of attraction. Sure, desire can diminish and even disappear, but sex doesn't have to. Having sex is a choice, not a feeling.

How Throw Blankets Affect Sexual Desire

Survival tip #38   It doesn't matter if your partner is George Clooney or George Costanza, your sexual desire is likely to wane over time. You may want sex less, and because of habituation, the same touch may feel less intense. But lust and attraction are not simply about how someone looks, or how they make you feel sexually. Attraction is based on external factors too, like whether your partner did something nice for you or respected your feelings. Attracting your partner and fueling their sexual desire might be as simple as folding a few throw blankets. Ask your partner what turns them on and off mentally, and then adapt.

Bonus Technology Tip

A divorce attorney I know is now working technology use into her pre- and postnuptial agreements. Why? Because she knows it's an issue. Couples fight over how and when a partner is on their phone. They feel rejected when a partner chooses to scroll through social media rather than have sex. They use their phones to have emotional affairs and correspond with paramours. They search each other's phones without permission. Before you even get married, you should talk about phone rules. What's acceptable and what's not? Don't wait until an issue arises to have this conversation. By then, it might be too late.

A MILLION LITTLE PIECES

My two most memorable stories about my grandma both involve potatoes.

The first is from when I was in my late twenties and I asked my grandma about her relationship with my grandfather. For reference, they divorced when their three daughters were adults, and he—Sam—died when I was only fourteen. I didn't know him well, but he was a watercolor artist, famous for his paintings of the pastural, midwestern landscape. When I was young, I would go to the house he shared with his new wife, and it was magical. This was due mostly to a feature I've seen in no home since: a built-in art gallery. All wood, with an open staircase and lots of light, the bottom two floors showcased his paintings, and the top floor was his studio, which connected to his master bedroom via a sliding glass door.

My mom had told me that Grandpa Sam was a great artist, but he hadn't been a wonderful husband. He spent long days away from the family on business in Chicago. It was rumored he had affairs. And he may have left my grandmother for his new, much younger wife. So when I asked my grandma about their relationship, I expected to hear a few of these details. Or maybe I just wanted some insight into her life as a wife and mother—the sides of her I never knew. Instead, she told me only one brief story.

"Your grandfather was somewhat difficult to get along with."

"In what way?" I asked.

"Well," she said, "whenever I made baked potatoes, he would ask me, 'Did you wash the potatoes?'"

"Uh-huh," I replied.

"He asked me every single time," she continued, "and I would always say, 'Yes, of course I washed the potatoes.'"

I remember looking at my grandma and wondering where this story was going.

And then she said, "It made me so mad that one day, you know what I thought? I thought—I'm going to wash those potatoes with soap!"

Not leave him. Not divorce him. Not tell him that his continual questions about the potatoes were undermining everything she did as a 1950s' wife. All that resentment my grandmother must have carried for not only the baked potatoes but also his countless acts against her, and my grandma's big plot for revenge was to wash his potatoes with soap. She was a saint.

This feeling I have about my grandmother's sainthood carries over into the next potato story. About five years before her actual death, my grandma had a big health scare. My aunts, mom, and I rushed to her hospital bed in New York. Her breath was shallow. She was weak but alert. My aunts were keeping vigil at her bedside. When I arrived with my mom, the doctor told us she had a leg infection—not life threatening—but she seemed to lack the will to live.

And yet she did possess the will to spill the beans.

Thinking she was dying, she began telling us about myriad issues she was having at her long-term care facility. Apparently, the nursing staff had been wheeling her into the main dayroom each day for entertainment, and one of the aides was reading the local newspaper to the captive residents. Nothing exciting, just local news, which was completely uninteresting to my grandmother, who had always been an avid reader and was just as sharp cognitively as she had been twenty years ago.

"None of us are even from this small town. I think they are reading it to us just so we fall asleep," she said.

"Do you fall asleep?" I asked.

"Yes, I do," she lamented.

But this wasn't the biggest offense.

"Sometimes the aides are not kind to me when I ask for my medication, and some days they tell me it isn't Wednesday when I know it is, and I end up missing my shower. And one time, they called us down for a group activity, and the activity was peeling potatoes."

Now I knew my grandmother's nursing home was not great. And the construction paper snowmen I was sent each year gave me some indication that her group activities were not stimulating. But this potato-peeling story made me want to run away from the hospital with my ailing grandma over my shoulder. I wanted her last years on this earth to be magical, not dismal. I wanted to rescue her.

But I didn't. I thought then and still believe now that I lacked the capacity to care for her in the way she needed. I could love her, but I couldn't *care* for her. It was the same for my mom and aunts. None of us had the capacity to provide her with the care and attention she needed as her body became fragile and her functionality declined.

My grandma lived three more years. She stayed in the same place, and silently endured the local newspaper reading and skipped showers. And with each passing day, a little bit of her light dimmed until she died in the potato-peeling care facility in 2017.

As I reflect on these stories today, I have two overwhelming feelings. One is that I now understand my grandma's baked potato story in a way that I couldn't when I was in my twenties. At the time, I thought that relationships always broke in one big piece, tearing two people apart violently because of one major event. Like when my grandpa left my grandma and began a life with his new wife. Now I realize that relationships often break in a million little pieces. When my grandma told me the baked potato story, she was picking up one little shard of her broken marriage and showing it to me. It was a symbol. She could have picked any of those pieces, but she chose the one with baked potatoes. The topic of the story was irrelevant. It was the sentiment, a broken marriage by a million little cuts, that she was trying to convey.

The other feeling I have is dread. My grandma's nursing home experience is my worst-case scenario. She was blind and alone in the last decade of her life, and crippled by the pain of arthritis and broken joints. Her last years were spent struggling to hear and feel the world around her in a variety of care homes where the people—residents and aides—were always changing. She spent most of her days listening to books on tape, participating in infantilized leisure activities, and being wheeled into a large dayroom to listen to someone read the local newspaper. My aunt and uncle would visit her every other afternoon for a couple of hours, but even that time for her was sometimes exhausting. In the end, I think she preferred to be alone, in her dark world, with the comfort of a soothing voice reading stories to her, taking her to a place far, far away.

You see, people don't always leave this world because of one major event. Instead, like my grandma, they often break in a million little pieces.

COLLAGEN PILLS WON'T SAVE YOU

As I've been aging myself, I've been really cognizant of the ways that people are trying to avoid their own worst-case scenarios. I've noticed that those around me—my friends and loved ones—are concentrating their antiaging efforts primarily toward three main areas: aesthetics, mobility, and disease prevention. I understand where these motives originate. Regarding the attempts at maintaining aesthetics, research by Buss and colleague David Schmitt has shown that physical attractiveness is one of the most important factors in mate selection, and the mainstream media sends us messages that promote beauty and youth.[1] Our obsession with all things beautiful even extends to disparate pay in the workplace, where attractive workers across many professions earn more than their unattractive counterparts. Economists refer to this as the "beauty premium." Although much of this imbalance can be attributed to other qualities that attractive people often have (e.g., intelligence, good health, and good personalities), the fact remains that an imbalance exists.[2] Perhaps as a result, we have a booming global antiaging market, which has been progressively growing over the last decade and is estimated to reach $83.2 billion by 2027.[3] From antiwrinkle cream to collagen pills, concoctions that help keep us looking and feeling young are ever growing in popularity.

Meanwhile, efforts toward maintaining mobility really resonate with me, especially after watching my grandmother become debilitated and increasingly immobile as she aged. While few of us will be able to maintain the physique and flexibility of Johanna Quaas, the famed ninety-one-year-old gymnast, most of us have the desire to be ambulatory and physically active in our elder years. This will allow us the most freedom in terms of where we live, how we spend our leisure time, and how we move about the world. Finally, although I mention it last, effort toward disease prevention seems to be the most common behavior I see practiced by those around me, particularly if I consider its multiple manifestations. From low-sugar diets to regular colonoscopies, as those around me age, they seem to choose one or multiple routes to maintaining their overall health.

Although those who purchase face creams and laser treatments are not always the same ones who are attending yoga classes and getting regular health screenings, from my limited observations, there does seem to be a good amount of overlap. According to the US Centers for Disease Control and Prevention (CDC), there may be overlap in general health behaviors as well. In its study of twenty-one-year-old and up people in the United States, 6.3 percent engaged in all five healthy behaviors measured that are linked with disease prevention (i.e., not smoking, regular exercise, minimal alcohol consumption, normal body mass index, and sleeping at least seven hours a night), and an additional 24.3 percent were almost completely compliant, engaging in four or five of the health-related behaviors.[4] Roughly speaking, this means that about one-third of the population is engaging in most or all of the behaviors deemed most important for disease prevention. And who are the people most likely to engage in all five health behaviors? Those sixty-five years and older.

Now this doesn't necessarily mean that as we get older, we get healthier. Sure, those over sixty-five are more likely to be retired than those in younger age groups and may have more time to devote to health-promoting habits. Or perhaps as they begin to face mortality for themselves along with their peers and family members, disease prevention is a more salient issue. Alternatively, it could be a period or cohort effect. That is, it could be something about the time period these seniors lived through or cohort in which they were born into that has always kept them healthier than the rest of us. If this is true, as our population ages, this promising statistic will go down. Particularly for metrics like maintaining normal body mass index, which was the health behavior that was least likely to be reported in the CDC survey (present among only 32.5 percent of the respondents), we see these period and cohort effects. As an example, a study in the United States showed that those born in 1975 had a 30 percent greater probability of being obese at age twenty-five than those born in 1955, and those born in 2002 had a 175 percent greater chance of being obese at twenty-five than those born in 1976.[5] This obesity epidemic is threatening our longevity. In fact, it is cited as one of the main reasons that life expectancy in the United States is going down for the first time in two centuries.[6]

Mortality statistics aside, seniors seem to be edging out the rest of the population in healthy habits. Except for one really important one: social engagement. A recent survey in the United States found that almost 25 percent of all people over the age of sixty-five are socially isolated, and more than one-third of those over forty-five report being lonely.[7] Plagued by socially ostracizing circumstances, like living alone, losing family or friends, and chronic illness, our elderly peers are often the loneliest and most socially isolated of all individuals in our population. Here I must distinguish, however, as the CDC has, between loneliness and social isolation.[8] Loneliness is an internal feeling that is not necessarily reflective of the outside world. Someone could be alone all the time but not feel lonely, and another person could be constantly in the company of others and still feel alone. Loneliness is subjective.

Social isolation, meanwhile, is defined by the lack of social connections and is an objective measure, reflective of the person's environment. Remember the social networking index scale I talked about in chapter 4? There's an abbreviated scale, called the Lubben Social Network Scale, that is recommended as a screener for older adult populations to measure whether they are socially isolated.[9] It contains only six questions, asking how many relatives the person sees or hears from at least once a month, how many they can talk with about private matters, and how many they feel close enough to contact if they need help. It asks the same three questions about friends. Individuals get points (on a scale) for every person they count in each category, and scores of twelve or lower are indicative of social isolation. Thus it's both the number and diversity of your *meaningful* contacts that matter, and more is better. Or rather, fewer is worse.

Social isolation among those fifty and older is a significant health threat, comparable to obesity and cigarette smoking in terms of increased mortality risk.[10] It is linked with increased rates of dementia, heart disease, and stroke. And loneliness—that subjective feeling—is associated with higher rates of depression, anxiety, and suicide.[11] For the second time in this book, I'll assert that social isolation can actually kill you. But how? And why?

GETTING IN THE (CRUMPLE) ZONE

This was exactly the question posed by James House, a sociology professor from the University of Michigan.[12] Apparently, having people around is good for you for many reasons. For one, being alone might spur a physiological response that is stressful to the body and, over time, detrimental to health. This might be counterintuitive to those of you who love the idea of alone time with a bath, book, and glass of wine. But try to imagine those solitary moments of respite stretching on indefinitely. And then put yourself into the body of my grandma, who needed help to even shower, could no longer see, and couldn't drink alcohol because if combined with her medications, it would have created a lethal cocktail. With these limitations in place, does the idea of being alone still spark joy?

Alternatively (or additionally), being with others may help buffer the stress of daily life or lower physiological arousal during extremely stressful events. This theory resonates with me because I like physics. There is something built into cars called the *crumple zone*. It's a safety feature specially designed to absorb impact—to give people more time and space during a head-on collision. Your meaningful people, they are your crumple zone. During extremely trying times, you need more of a buffer, and this is why people rush to your side during a tragedy. But daily life can also get us down. It can wear us in ways that have a cumulative impact. And those times you spend with friends and family out to dinner, watching a game, or around the pub are your crumple zone. They create space and time between you and life's daily stressors.

House provided two additional reasons for which social interaction might help us live longer, and they both relate to positive support. First, our meaningful people not only provide tangible support but also are a potential source of helpful social pressure. They might encourage us to eat healthier, exercise, and engage in life-preserving behaviors like taking medication. They might discourage detrimental behaviors such as an excess of drinking and drug abuse too. Second, through their networks, they might facilitate our acquisition of resources that help keep us healthy, like jobs and medical referrals. Here I think of my friend Julie, who invites me to hot yoga every Saturday even though I turn her down 75 percent of the time, is willing to give up bread with me when I need to fit into a small dress even though she's a size zero, and consults her network of school counselors when I have questions about my sons' education. She's there for me. She inspires me. She propels me. But even if someone doesn't have friends and family members like Julie who provide this type of positive support, something (even a subpar something) appears to be better than nothing.

And that brings me back to my grandma. Long before my grandma entered the care facility, she divorced her second husband, Walter. I remember little about him either, except that he collected pennies, read *National Geographic*, and spent almost all his time in the basement when we visited. When my grandma told me she was divorcing him when she was in her seventies and almost completely blind, I felt the weight of her decision. He had been her link to the outside world. He bought the groceries and cleaned the house. He drove her each day to the senior site where she volunteered to roll silverware in paper napkins for the daily lunch service there (the only job she could do with her vision so compromised). Nonetheless, *she* left *him*. Rather than spend one more minute with Walter, she chose a life of severe isolation, where her daily contact was nurses' aides and the other residents, many of whom were nonambulatory or had dementia. But these contacts were not stable. The average annual turnover rate for aides in nursing homes is 59 percent, and the lowest-quality nursing homes have the highest turnover rates.[13] Not surprisingly, mortality among residents is equally dismal. A recent study of nursing home residents in Norway showed that about one-third of them died each year, and the median survival was 2.2 years, which is similar to the 2.1- to 2.6-year survival rate reported in other countries.[14] When my grandma finally died at ninety-three, she had outlived almost everyone she had ever known.

I say all of this to highlight a simple fact. When you think about growing old, you might think about your children moving away and not visiting often. You might think about having to cope with the death of your friends and family members. You might even think about facing disease. But something few of us think about is the stress of daily loss. Loss of familiar people. Loss of freedom. Loss of abilities. My grandma was not only alone. She was alone and aware that every day she was losing a piece of her fragile world.

IN SICKNESS AND IN HEALTH

I'm not sure why my grandma chose to leave Walter. I sometimes wonder whether her blindness created additional stress on an already-strained marriage. I do know that illness and compromised functionality can be greatly disruptive to a couple. There are many issues that emerge throughout life that can challenge a couple and help them grow. Financial strains, infidelity, and disagreement on major issues—couples can often overcome these issues and emerge stronger. The stress of a partner's illness, however, is frequently devastating. Maybe this is because the healthy partner has to pick up additional tasks or provide additional emotional support to maintain life, and this is draining. Yet it might also be because the ill partner changes in ways that makes the relationship dynamic different.

When a person suffers a major health crisis or gets diagnosed with a chronic disease, it can be extremely discombobulating. One day they were just walking through this world as a person without a diagnosis. The next day, the disease or crisis becomes part of their identity. No longer are they simply Paula, Ryan, or Lucy. Now they are Paula, the patient with rheumatoid arthritis, or Ryan, the patient with Parkinson's disease, or Lucy, the patient undergoing breast cancer treatment. The correct diagnosis might take some time to figure out, but once it is determined, it is delivered in an instant. Cancer, diabetes, dementia, or multiple sclerosis—what may have begun as a few nagging symptoms changes immediately into a planned course of treatment for a potentially life-threatening disease. And that's if they have a chance of survival.

Once diagnosed, people can exhibit a wide range of emotions. Some accept their diagnosis and try to immediately adapt as well as they can into this new phase of life—one that may now be monopolized by doctors' appointments and hospital stays. Others, however, have a much more difficult time. They might become information addicts, obsessed with reading everything they can find about the disease, its treatment, and prognosis. They might suffer anxiety and depression, as there is a strong relationship between chronic disease and depression across a large range of conditions. As an example, of those who have a heart attack, 40 to 65 percent experience depression, and of those diagnosed with diabetes, 25 percent experience depression.[15] As might be expected, rates of depression seem to correspond with disease severity and how much the disease disrupts the patient's life.

So even when the illness does not cause a person to socially isolate because of restricted mobility or other types of functional disability, the emotions associated with coping with the illness might make them withdraw from family members or other social connections. In a time when they might need intimacy more than ever, they might be especially hard to reach. Others might shift back and forth between different emotional states, including guilt and grief. Shock, acceptance, and despair—all of the same emotions that people might have when told that their life may be ending—might also be present, in unpredictable phases, when people are diagnosed with a chronic or life-threatening disease. And all of these reactions are normal.

You see, there is no one way to respond to a life-changing diagnosis or health crisis. People respond in drastically different ways to different illnesses, but even with the same diagnosis, reactions vary substantially. Yet in terms of health benefits, some responses are more helpful than others. In one seminal study, Canadian researchers followed women who had been diagnosed with breast cancer to examine how their coping responses at the time of their treatment planning related to their psychological adjustment three years later.[16] Their main finding was that those who resigned themselves to the disease or avoided the realities of their diagnosis, and showed signs of depression while planning their course of treatment, were significantly worse off psychologically three years later. Notably, this resignation to their fate, and feeling out of control and hopeless, was an especially strong predictor of these women's later mental state above and beyond their level of depression. Hope, it seems, is paramount to coping.

One thing that can help people cope? Friends. A childhood friend of mine is a breast cancer survivor. The incidence of breast cancer has been rising at staggering rates over the past few decades, as nations become more industrialized. The rates are so high, in fact, that in the United States, it is estimated that one in eight women (12 percent) will have breast cancer in her lifetime.[17] At my friend's final chemotherapy treatment last year, her many friends who had been accompanying her to each of her treatments all showed up to the hospital with balloons in hand and scarves on their heads. They were there cheering in solidarity as she took her last steps in beating a devastating, life-threatening illness and hearing those long-awaited words "no evidence of disease." Research supports these friendly efforts. In a study of 393 Indian women, those who had more support from friends were less likely to exhibit hopelessness.[18] And those who had support from significant others were more likely to have a strong fighting spirit. Support from those we love is key to coping. They help us get through the depressing times, the lonely times, the times that challenge our spirit and make us want to go cry alone in a corner.

And yet that support can sometimes come with a significant burden. In a study of 2,701 marriages from the RAND Health and Retirement study, which includes a representative group of people in the United States over fifty years of age, researchers found that a spouse's illness earlier in the marriage has a significant, detrimental effect on marital health. Notably, a cancer diagnosis in either the wife or husband was not predictive of divorce—a finding that contradicted previous research. Still, the onset of both wives' and husbands' heart or stroke issues earlier in their relationship were predictive of being widowed later on, and when a wife had lung problems or a stroke, it increased the likelihood of divorce. These results illustrate a trend that's difficult to reconcile. On the one hand, support from significant others can be uplifting and help fuel the fighting spirit so imperative to maintaining good mental health. On the other hand, it can damage a marriage, perhaps irreparably. Although it is common for people to vow to stick with their partners in sickness and health, the "in sickness" part is often a difficult vow to keep.

"I FELT SUCH RELIEF I WAS NOT ALONE"

Luckily, we don't have to rely solely on our existing networks for social support. One of the most helpful and burgeoning markets of online support involves those related to health conditions. For almost every type of disease or affliction you can imagine, you can find a support group on the internet. Relevant to my previous example, with a quick internet search you can easily find a discussion board that links you to other patients with breast cancer. You can connect with them to share your woes and triumphs. You can exchange treatment regimes and discuss different ways to cope with side effects. You can openly share the stresses of your disease, and how it has affected your marriage and family. And those listening to you, unlike many of your friends and family members, may be able to empathize better than those in your social circle because they have walked in your shoes. They understand you in a way that others may not.

But that's not all that these online networks have to offer; they may be sharing their patient data too. If you haven't heard of the website patientslikeme.com, it's time you did. Patients-LikeMe (PLM) is an online network where patients can connect with others coping with the same condition. Caregivers can join as well on behalf of the patient and are instructed to create posts as if they were the patient. As of July 2021, the website boasted more than twenty-eight hundred specific conditions around which more than six hundred thousand members gathered to discuss anything from disease progression, to treatment regimes, to their current mood and symptoms. Researchers have shown these online networks to be beneficial, providing not only social support but also encouragement and education related to disease management.[19]

According to one of its members, there are other benefits too, including relief from the lonely journey of illness. Craig, a PLM member, offered a testimonial for the website that proclaims, "When diagnosed, I felt overwhelmed and alone in the world. I joined PatientsLikeMe and found there were thousands of others with my condition. I felt such relief I was not alone." For patients like Craig, perhaps connection was an important part of his own disease management. Yet the benefits of connection extend far past Craig and his disease. PatientsLikeMe is a prototype for health data information sharing.

One of the activities patients engage in on this website is sharing their own health data. Although they do not share information like their name and social security number, they share highly personal medical information such as demographic information, lab results, and side effects. They share it as a sort of longitudinal record—one that others can track and use to compare against their own experience. One can see easily how this might be beneficial to a person suffering from Parkinson's disease, for example, as it might show them the best and worst outcomes that are associated with different disease states and treatments.

Undeniably, these data from more than half a million patients is valuable to the research community too. Patients who enroll in PLM (which is a for-profit organization) give permission for their general data to be shared with researchers, who may, for instance, use data from these participants as a control group for an experimental trial. The website is also used as a recruitment vehicle for researchers interested in specific patients for clinical trials, including virtual trials. In a survey of PLM members,[20] community members were mainly in favor of data sharing. PLM members also cite numerous benefits of this data sharing, including helping their doctors with decision-making concerning their condition, and facilitating the development of new treatments and therapies.

On their website, PLM indicates that the organization makes money through these research collaborations, in part by sharing anonymized health data that its members provide in their online communities. The members curate their own open-source medical records, and these records are invaluable to research scientists who are studying disease. Why? Because this is usually heavily guarded, private information. According to existing regulations in most countries, your health information is available to you and your provider (who may share your unidentified data with others). But unless you give permission for it to be shared with another person or entity, it's available to no one else on this earth.

HOW AMAZON, APPLE, AND IBM ARE CHANGING THE WORLD (EVEN MORE THAN WE THINK)

In our current world of health care, we have little interoperability or data sharing across health systems.[21] Even in countries with universal health care like China, interoperability is not commonplace because hospitals do not use one common system for the maintenance of electronic health records. In fact, perhaps by design, those systems are often incompatible. It is time consuming and arduous to transfer patient data from one system to another. This is why you sometimes have to have a scan at one clinic, and then you have to pick up your results and physically deliver them to your doctor, or sign a waiver for them to be faxed over. But what if these barriers did not exist? What if your doctor could look in your records to see the results and notes from all of the tests, scans, and visits you've ever had? Wouldn't this give them a more comprehensive picture of what was happening? And taking it one step further, what if they could then take those data and plot them against the data from people all over the world, matching you by factors like age, weight, and preexisting conditions, to see how your test results compare to those of other patients with similar issues? Imagine how that might aid not only diagnosis but also patient care and beyond. It could be revolutionary.

This is the reason why in some regions of the world, like the European Union, policy makers are making extreme efforts toward cross-border interoperability. It makes sense that they would lead the way, as universal health care is commonplace within Europe, and citizens are able to travel, work, and live within the countries of the European Union. In a recent survey of infrastructure, the EU division of the Healthcare Information and Management Systems Society found that some countries like Spain and England (which officially seceded from the European Union in 2020) have good levels of operability within and across their health care systems; in both cases 89 percent of their hospitals have systems that can share data across multiple locations.[22] England also has already made substantial moves in this direction, implementing a patient data information sharing program in several regions across the country.[23] In contrast, the Healthcare Information and Management Systems Society survey showed that Germany's interoperability lagged far behind; more than half of the hospitals (especially the smaller ones) reported that they had no interoperability at all, and 45 percent had no plans for developing an interoperable system.

Despite this, tech companies are still moving forward, creating the potential infrastructure through which medical data can one day be exchanged. One of those is Apple. The iPhone Health Records application, which was launched in 2018, allows patients from more than five hundred health care organizations, including Veterans Affairs, to access their health records on their phones. This information is available not only to patients. As of June 2018, Apple also opened the application up to developers so that when individuals using the application choose to do so, they can share their health record data with third-party applications. Users of iPhones can download the Apple Research Application as well in order to participate in research studies related to issues like heart issues, hearing, and women's health. Other companies are creating products geared toward health systems. For instance, with its cloud-based service iConnect, IBM's Watson Health is pushing the industry toward interoperability for image exchange.[24] Providers using this system would have access to not only current and historical patient images but also be able to coordinate with other providers within a single system. In fact, IBM is one of the technology powerhouses—along with Apple, Amazon, Microsoft, Salesforce, and other US companies—that signed a pledge in 2018 to try to develop tools and standards that will help make moves toward interoperability along with safe and efficient patient data exchange.[25]

Not surprisingly, Amazon has extended its reach to all sorts of data in the form of the AWS Data Exchange. Once a seller of books, Amazon's AWS marketplace has now expanded the world's most famous seller into the highly profitable market of data.[26] Boasting data from a variety of different types of customers including Dun & Bradstreet, which has data on more than 330 million businesses worldwide, and Foursquare, which tracks consumer location data and information from more than 60 million commercial venues, the AWS Data Exchange offers a glimpse of the future of data as a commodity. The possibilities of data exchange are endless. And health care entities have gotten in on the game too. As an illustration, within the AWS marketplace, Change Healthcare apparently provides access to a database of more than 14 billion health care transactions. Access to many of these databases is associated with a cost. It is a marketplace after all. As of 2021, though, there are more than a thousand data sources that offer their databases for free. Some of these are related to COVID-19. As an example, Tableau, in conjunction with Salesforce and MuleSoft, is offering coronavirus case and death data from trusted sources like the *New York Times* and European CDC.[27]

Organizations have also created standards to help protect patient data. In the area of health information technology, there are currently more than forty standard development organizations that have been accredited by the American National Standards Institute or International Organization for Standardization.[28] These standard development organizations, usually comprised of clinicians, health care administrators, informaticists, and information technology specialists, provide structural guidance for everything from the terminology used for health concepts, to the way data will be transported, to the privacy and security standards that will dictate how the data are collected, transferred, and stored.[29]

In short, the technical, ethical, and clinical infrastructure is being set up globally, priming the world for unprecedented data exchange that will improve precision in patient care. In a historic move in 2020, patients were granted more rights related their own health data, thereby potentially improving interoperability as well as giving physicians access to wide databases of patient outcomes for different types of diseases and treatments.[30] There are also companies such as US-based Hu-manity.co that give patients the right to sell their own data. In the absence of a global agreement about who owns patient data, Hu-manity.co is pushing our boundaries on ethical human rights related to digital data ownership. Our data are already monetized. In exchange for "free services," social network sites, for example, sell our user data. Yet Hu-manity.co founder and CEO Richie Etwaru wants to bring purposeful decision-making and ownership to this data exchange, empowering individuals to decide to whom to sell their data and for how much. It's a novel idea based on transactional transparency and is still in its early stages of development. Although it certainly applies to highly protected and valued patient data, if ideas like this gain traction, there may be millions of data points that can be monetized throughout a person's lifetime.

Considering these historic changes in the way humans can and will use data, alongside the popularity of websites like PLM, it is obvious that we are heading in a direction of easier, more transparent health data sharing. Although this undoubtedly will spur ethical discussions about the privacy of health information and policy changes to lower the potential for that information to be used for harm, it is a promising step for the future of medicine too. Connecting humans in intimate ways will involve connecting them not only as individuals (e.g., finding fellow moms whose children have cystic fibrosis through blogs or online groups) but also as masses of data points in an interconnected network of health-related information.

In my mind, I see a promising future of medicine. Health systems will be operating as a single system, providing a large web of interconnected personal health data—the most personal data an individual might ever possess. And these data will be mined by sophisticated computer programs to help diagnose and treat individuals all over the world, thus increasing safety and precision. This possibility is enthralling, as there might be nothing more intimate than helping another human survive.

Maybe the robot doctor I envisioned at the start of this book will never be in a position to give a devastating diagnosis. Health care and digital ethicists might protect all of us from that reality. Instead, my human doctor will pull up my images on a clear screen, showing me my course of treatment and expected prognosis, mapped against thousands of other data points from patients like me. And in my darkest moment, they will hold my hand and comfort me—a touching reminder of the intimacy that binds us all together as humans.

A FINAL NOTE: MY LAST MOMENTS WITH SOPHIA

When I was in Hong Kong filming the pilot with Sophia, I became friends with my British producer. He and I both suffered terrible jet lag for the duration of our trip, and we would regularly send each other messages in the early morning as we waited for 5:00 a.m. breakfast. Dim sum for me; eggs and toast or fruit for him. During that week, we talked about everything from our families, to the future of robotics, to the Netflix documentary on Italian porn star Rocco Siffredi. We developed the scripts for the "dates" and discussed how to troubleshoot the issues we were having during filming. We made small talk as we traveled back and forth in a van from our dark, makeshift studio back to our gleaming, modern hotel. On set, he'd look at me curiously, like I was a puzzle he was trying to figure out, and then look back through the cameraperson's lens and tell them to make an adjustment.

I know now that he was trying to get to the heart of me. And on the last day of filming, he did.

"Do you think that these robots could really make a difference?" he asked.

He asked the question from a place off camera, and I was sitting on a metal chair in the middle of the room. The studio was dark, and a single light made a circle around me.

"Yes, of course," I replied. Stoic. Scientific. Logical. "How?"

"Well maybe for people like you and me, robots like Sophia won't make much of a difference. We're lucky. We have love and friendship and lives full of adventure. But these robots, they could really be helpful for people who feel lonely."

"Like who?" he continued, looking at me from his dark corner.

He had been teasing me all week that we would get me to cry on camera, and I assured him I wouldn't. But I felt my voice catch, and I knew instantly that all my assurances would crumble. I said the rest through tears.

"There are people on this earth who are desperately lonely. They are starving for love and attention. They need someone to talk to, and someone to listen to them. I think about Sophia and wonder how much better my grandma's life would have been if she'd had her by her side. She was blind and alone when she died. Sophia could have made her every day infinitely better." Then my professional facade gave way to a flood of me as I said my final piece:

"From the start of life to its bitter end, everyone deserves to be loved."

HOW TO SURVIVE GROWING OLD

A Million Little Pieces

Survival tip #39 It is rarely one fracture that makes something break. Whether it be relationships or people, be aware of the ways your daily behaviors might be breaking them down.

If you want something to survive, you must nurture it. Every day. Every moment. If you have the choice between making something better or making it worse, stop asking about the potatoes because you're wrong.

Collagen Pills Won't Save You

Survival tip #40 Our efforts to safeguard ourselves as we grow old should not be limited to only our physical health. We also need to protect and enrich our social health. Invest just as much time (or more) in creating a diverse and rich social network as you do engaging in disease prevention. These meaningful people may one day enrich or extend your life in important ways.

Getting in the (Crumple) Zone

Survival tip #41 It is a human tendency to rally around others when they experience trauma. It's what our culture teaches us, and it feels good to be there for the people when they are facing tough times. Still, tough times are not always obvious. And people may need you to show up for them just as much on their so-so days as they do on their darkest ones. Just show up. It's always a winning move.

In Sickness and in Health

Survival tip #42 Illness and health crises are often painful and emotionally taxing to cope with. But they are also extremely isolating and stressful for couples and families. When someone becomes ill, it's important to extend your love and support them, even if they withdraw. Offer to drive them to appointments. Drop off cookies at their door. Attend their chemotherapy sessions. It will give them hope and enhance their fighting spirit. This is crucial for their mental health.

"I Felt Such Relief I Was Not Alone"

Survival tip #43 If your off-line world is deficient in the support you need, don't hesitate to look online for people with whom to connect. There are thousands, if not millions, of online forums that connect patients and caregivers to people dealing with the same health issues you are grappling with. Don't hesitate to find those people, and share your woes and triumphs. When life is off-kilter, knowing you're not the only one who is dealing with a condition or crisis may be all you need to regain your balance.

How Amazon, Apple, and IBM Are Changing the World (Even more than We Think)

Survival tip #44 Connection comes in many forms. We are connected to individuals and groups. But we are connected through a web of experiences and actions to all the other humans on this earth too. Big tech is making moves to strengthen these connections in ways that give us more access to information and the ability to effect change in people's lives through the sharing of our digital data. Pretty soon, the decision to share your data and create networks of interconnected information that might transform everything from transportation to health care will be in your hands. Be prepared to act. Quickly. Because as you know, technology moves fast.

A Final Note: My Last Moments with Sophia

Survival tip #45 By the end of my lifetime, I expect to see drastic changes in the robotics industry. My dreams of companions for the elderly and otherwise lonely may become a reality as tech companies push forward with innovations in the fields of AI and animatronics. I nonetheless still have so many questions. What laws and guidelines do we need to keep humans (and robots) safe and healthy? How might the integration of robots into our lives shape human-to-human interactions? And will we one day be able to love robots in the same way we love each other? We will need to approach this future with these ethical and developmental questions in mind.

Bonus Technology Tip

As we age, we need to think about the ways in which we can use technology to keep us healthy and connected to others. Yet the advances we've seen in technology in the last decade foretell a fast-paced future in which it will be almost impossible to keep up with the latest trends. Depending on where you are in your life, it might be tough to imagine a time when you will be out of touch with the technological innovations of the era. It will come for all of us, however. And when that day comes, try your best to be flexible and be prepared to shift quickly.

Afterword

I wish I could tie this up with a lavish bow, and declare the subject closed and problem fixed. But that's not the point of this book. The point was to draw your attention to the myriad issues we are confronting as humans in modern society and get you to really *think* about them, perhaps in ways you've never thought about them before. Meanwhile, the tips I provided along the way are simply my thoughts on how to make life a little bit better. They are not prescriptions or solutions. They do reflect, however, some deep thinking I've done on topics that are really important to me and, I hope, you.

That said, knowing that each of you may want your time investment in this book to end with a concrete takeaway, I offer this final exercise. I'd like you to focus on one of the chapters in this book that means the most to you in the current phase of your life. Then via a framework commonly used for self-improvement, answer these three questions:

1. What's working for me?
2. What's not working for me?
3. What am I missing?

You can do this with each of the topics. You can even ask the people in your life to do the same. But the most important action you can take is to hold yourself accountable to continue to do the things that work, ditch the things that don't, and invite into your life through intention those things you are still missing.

As for me, I have one last reflection that I believe will help me tidy up this book from my headspace. About fifteen years ago, I took a vacation by myself. At the time, my husband and I were living in Jyväskylä, a small city in northern Finland. I needed some time away to think through some problems we were having in our relationship. In truth, I wanted to run away. The problems we were facing were serious, and as a daughter of a rolling stone, I had no model for "working it out." So I flew to Tenerife, one of the Canary Islands, thinking that a book and beach were all I needed to reflect and heal. And maybe *move on*. I realized quickly, though, that my healing wouldn't happen in isolation. In fact, I felt as lonely as I'd ever felt in my time on this earth. And that loneliness smothered me like a dark, thick blanket.

After two days of speaking to no other human but the receptionist, I changed my plane ticket and flew back home to Finland. Instead of sitting on a beach alone, my husband and I walked nightly through serene, snow-covered forests. We cried together. We held hands. We spent restless nights having difficult conversations. My healthy and socially adept partner helped me realize that I am not an island. I need people. And

I need love. And when things get tough, I shouldn't push either of those things away. It's basic really. Since that day, I've made every effort to love both unabashedly and with commitment. And through this book, I extend this love to you.

Dear book, "You served me well, now goodbye."

Acknowledgments

To my mom, who taught me how to love. To my husband, who taught me that love can endure. And to my sons, Tyler and Noah, who taught me that even when you think your heart is full, there is always room for more.

Notes

INTRODUCTION

1. Andreas Kaplan and Michael Haenlein, "Siri, Siri, in My Hand: Who's the Fairest in the Land? On the Interpretations, Illustrations, and Implications of Artificial Intelligence," *Business Horizons* 62, no. 1 (January–February 2019): 15–25, https://doi.org/10.1016/j.bushor.2018.08.004.

2. Mike Murphy, "Replika: This App Is Trying to Replicate You," *Quartz*, last modified August 29, 2019, https://qz.com/1698337/replika-this-app-is-trying-to-replicate-you.

3. Jean Twenge, Brian Spitzberg, and W. Keith Campbell, "Less In-Person Social Interaction with Peers among U.S. Adolescents in the 21st Century and Links to Loneliness," *Journal of Social and Personal Relationships* 36, no. 6 (June 2019): 1892–1913, doi:10.1177/0265407519836170.

4. John F. Helliwell, Richard Layard, and Jeffrey D. Sachs, eds., *World Happiness Report 2018* (New York: Sustainable Development Solutions Network, 2018), 147.

5. Robert L. Kahn and Toni C. Antonucci, "Convoys over the Life Course: Attachment, Roles, and Social Support," in *Life Span Development and Behavior 3*, ed. Paul B. Baltes and Orville G. Brim (New York: Academic Press, 1979), 253–286.

6. Monica Anderson and Jingjing Jiang, "Teens, Social Media and Technology 2018," Pew Research Center, May 31, 2018, https://www.pewresearch.org/internet/2018/05/31/teens-social-media-technology-2018.

7. Jean Twenge, Ryne Sherman, and Brooke Wells, "Declines in Sexual Frequency among American Adults, 1989–2014," *Archives of Sexual Behavior* 46, no. 8 (March 2017): 2389–2401, doi:10.1007/s10508-017-0953-1.

8. Stuart Brody, "The Relative Health Benefits of Different Sexual Activities," *Journal of Sexual Medicine* 7, no. 4 (January 2010): 1336–1361, doi:10.1111/j.1743–6109.2009.01677.x.

9. Andrea L. Meltzer, Anastasia Makhanova, Lindsey L. Hicks, Juliana E. French, James K. McNulty, and Thomas N. Bradbury, "Quantifying the Sexual Afterglow: The Lingering Benefits of Sex and Their Implications for Pair-Bonded Relationships," *Psychological Science* 28, no. 5 (March 2017): 587–598, doi:10.1177/0956797617691361.

10. Claudia Schmiedeberg, Bernadette Huyer-May, Laura Castiglioni, and Matthew D. Johnson, "The More the Better? How Sex Contributes to Life Satisfaction," *Archives of Sexual Behavior* 46, no. 2 (February 2017): 465–473, doi:10.1007/s10508-016-0843-y.

11. Amy Muise, Ulrich Schimmack, and Emily Impett, "Sexual Frequency Predicts Greater Well-Being, but More Is Not Always Better," *Social Psychological and Personality Science* 7, no. 4 (May 2016): 295–302, doi:10.1177/1948550615616462.

12. George Loewenstein, Tamar Krishnamurti, Jessica Kopsic, and Daniel McDonald, "Does Increased Sexual Frequency Enhance Happiness?," *Journal of Economic Behavior and Organization* 116 (August 2015): 206–218, https://doi.org/10.1016/j.jebo.2015.04.021.

13. Bernie Zilbergeld and Carol Rinkleib Ellison, "Desire Discrepancies and Arousal Problems in Sex Therapy," in *Principles and Practice of Sex Therapy*, ed. Sandra R. Leiblum and Lawrence A. Pervin (New York: Guilford Press, 1980), 65–106.

14. Peter Ueda, Catherine H. Mercer, Cyrus Ghaznavi, and Debby Herbenick, "Trends in Frequency of Sexual Activity and Number of Sexual Partners among Adults Aged 18 to 44 Years in the US, 2000–2018," *JAMA Network Open* 3, no. 6 (June 2020): e203833, doi:10.1001/jamanetworkopen.2020.3833.

15. Kyle J. Foreman, Neal Marquez, Andrew Dolgert, Kai Fukutaki, Nancy Fullman, Madeline McGaughey, Martin A. Pletcher, et al., "Forecasting Life

Expectancy, Years of Life Lost, and All-Cause and Cause-Specific Mortality for 250 Causes of Death: Reference and Alternative Scenarios for 2016–40 for 195 Countries and Territories," *Lancet* 392, no. 10159 (November 2018): 2052–2090, doi:org/10.1016/S0140–6736(18)31694–5.

CHAPTER 1

1. Daniel Dunford, Becky Dale, Nassos Stylianou, Ed Lowther, Maryam Ahmed, and Irene de la Torre Arenas, "Coronavirus: The World in Lockdown in Maps and Charts," *BBC*, last modified April 7, 2020, https://www.bbc.com/news/world-52103747.

2. "Oxford COVID-19 Government Response Tracker," Blavatnik School of Government, University of Oxford, https://covidtracker.bsg.ox.ac.uk.

3. Emma Beswick, "Sweden's Coronavirus Strategy: Has 'Culture of Conformity' Saved the Country from COVID Fatigue?," *Euronews*, last modified October 19, 2020, https://www.euronews.com/2020/10/19/has-sweden-s-coronavirus-strategy-helped-it-avoid-pandemic-fatigue.

4. Sarah Mervosh, Giulia McDonnell Nieto del Rio, and Neil MacFarquhar, "'Numb' and 'Heartbroken,' the U.S. Confronts Record Virus Deaths," *New York Times*, last modified February 21, 2021, https://www.nytimes.com/2020/12/10/us/coronavirus-death-record.html.

5. Al Goodman, Laura Perez Maestro, Ingrid Formanek, Max Ramsay, and Ivana Kottasová, "Spain Turns Ice Rink into a Morgue as Coronavirus Deaths Pile Up," *CNN*, last modified March 24, 2020, https://www.cnn.com/2020/03/24/europe/spain-ice-rink-morgue-coronavirus-intl/index.html.

6. Erik Erikson, *Childhood and Society* (New York: W. W. Norton and Company, 1950).

7. Abby Vesoulis, "'If We Had a Panic Button, We'd Be Hitting It.' Women Are Exiting the Labor Force En Masse—and That's Bad for Everyone," *Time*, October 17, 2020, https://time.com/5900583/women-workforce-economy-covid.

8. James Crowley, "Hinge Introduces 'Date from Home' Feature So Users Can Date and Social Distance with Ease," *Newsweek*, April 7, 2020, https://www.newsweek.com/hinge-date-home-feature-virtual-dates-1495615.

9. Natsumi Sawada, Emilie Auger, and John Lydon, "Activation of the Behavioral Immune System: Putting the Brakes on Affiliation," *Personality and Social Psychology Bulletin* 44, no. 2 (October 2017): 224–237, doi:10.1177/0146167217736046.

10. Emma Ailes, "'Covid Ended Our Marriage': The Couples Who Split in the Pandemic," *BBC*, December 3, 2020, https://www.bbc.com/news/world-55146909.

11. Alexis Benveniste, "Taylor Swift Released Two Albums in 2020. Most of Us Are Just Trying to Get By," *CNN*, last modified December 11, 2020, https://www.cnn.com/2020/12/10/success/taylor-swift-albums-productivity/index.html.

12. Benjamin Jones, E. J. Reedy, and Bruce A. Weinberg, "Age and Scientific Genius" (working paper, National Bureau of Economic Research, Cambridge, MA, January 2014), https://www2.nber.org/papers/w19866.pdf.

13. Kim Kavin, "Dog Adoptions and Sales Soar during the Pandemic," *Washington Post*, August 12, 2020, https://www.washingtonpost.com/nation/2020/08/12/adoptions-dogs-coronavirus.

14. Linda Handlin, Eva Hydbring Sandberg, Anne Nilsson, and Mikael Ejdeback, "Short-Term Interaction between Dogs and Their Owners: Effects on Oxytocin, Cortisol, Insulin and Heart Rate—An Exploratory Study," *Anthrozoös* 24, no. 3 (April 2015): 301–315, doi:10.2752/175303711X13045914865385.

15. Erika Friedmann, Aaron Honori Katcher, James J. Lynch, and Sue Ann Thomas, "Animal Companions and One-Year Survival of Patients after Discharge from a Coronary Care Unit," *Public Health Reports* 95, no. 4 (July 1980): 307–312.

16. G. Adee A. Schoon, Danielle De Jonge, and Patrick Hilverink, "How Dogs Learn to Detect Colon Cancer—Optimizing the Use of Training Aids," *Journal of Veterinary Behavior: Clinical Applications and Research* 35 (January 2020): 38–44, doi:10.1016/j.jveb.2019.10.006.

17. Alan M. Beck and N. Marshall Meyers, "Health Enhancement and Companion Animal Ownership," *Annual Review of Public Health* 17 (1996): 247–257, doi:10.1146/annurev.pu.17.050196.001335.

18. Emily W. Flanagan, Robbie A. Beyl, S. Nicole Fearnbach, Abby D. Altazan, Corby K. Martin, and Leanne M. Redman, "The Impact of COVID-19

Stay-at-Home Orders on Health Behaviors in Adults," *Obesity (Silver Spring)* 29, no. 2 (February 2021): 438–445, doi:10.1002/oby.23066; Michael S. Pollard, Joan S. Tucker, and Harold D. Green Jr., "Changes in Adult Alcohol Use and Consequences during the COVID-19 Pandemic in the US," *JAMA Network Open* 3, no. 9 (September 29, 2020): e2022942, doi:10.1001/jamanetworkopen.2020.22942.

19. Aaron Tilley, "Zoom's Pandemic-Fueled Boom Continues," *Wall Street Journal*, last modified November 30, 2020, https://www.wsj.com/articles/zooms-pandemic-fueled-boom-continues-11606772231.

20. Rachel Sandler, "Here's When Major Companies Plan to Go Back to the Office," *Forbes*, August 27, 2020, https://www.forbes.com/sites/rachelsandler/2020/08/27/heres-when-major-companies-plan-to-go-back-to-the-office/?sh=3e71941f361c.

21. International Monetary Fund, *World Economic Outlook: A Long and Difficult Ascent*, October 2020, https://www.imf.org/en/Publications/WEO/Issues/2020/09/30/world-economic-outlook-october-2020.

22. International Monetary Fund, *World Economic Outlook: Managing Divergent Recoveries*, April 2021, https://www.imf.org/en/Publications/WEO/Issues/2021/03/23/world-economic-outlook-april-2021.

23. Janette Neuwahl Tannen, "Pandemic Spurs a Burst of Technology Innovation," News@TheU, University of Miami, August 18, 2020, https://news.miami.edu/stories/2020/08/pandemic-spurs-a-burst-of-technology-innovation.html.

24. Juulia T. Suvilehto, Enrico Glerean, Robin I. M. Dunbar, Riitta Hari, and Lauri Nummenmaa, "Topography of Social Touching Depends on Emotional Bonds between Humans," *Proceedings of the National Academy of Sciences* 112, no. 45 (November 10, 2015): 13811–13816, https://doi.org/10.1073/pnas.1519231112.

25. Klaus Linde, Kirsten Sigterman, Levente Kriston, Gerta Rücker, Susanne Jamil, Karin Meissner, and Antonius Schneider, "Effectiveness of Psychological Treatments for Depressive Disorders in Primary Care: Systematic Review and Meta-Analysis," *Annals of Family Medicine* 13, no. 1 (January 2015): 56–68, doi:10.1370/afm.1719.

26. Johan Ormel, Philip Spinhoven, Ymkje Anna de Vries, Angélique O. J. Cramer, Greg J. Siegle, Claudi L. H. Bockting, and Steven D. Hollon,

"The Antidepressant Standoff: Why It Continues and How to Resolve It," *Psychological Medicine* 50, no. 2 (November 29, 2019): 177–186, doi:10.1017/S0033291719003295.

27. Christopher A. Moyer, James Rounds, and James W. Hannum, "A Meta-Analysis of Massage Therapy Research," *Psychological Bulletin* 130, no. 1 (January 2004): 3–18, doi:10.1037/0033–2909.130.1.3.

28. Mark Hyman Rapaport, Pamela Schettler, Erika R. Larson, Sherry A. Edwards, Boadie W. Dunlop, Jeffrey J. Rakofsky, and Becky Kinkead, "Acute Swedish Massage Monotherapy Successfully Remediates Symptoms of Generalized Anxiety Disorder: A Proof-of-Concept, Randomized Controlled Study," *Journal of Clinical Psychiatry* 77, no. 7 (July 2016): e883–e891, doi:10.4088/JCP.15m10151.

29. "Massage Therapists," Occupational Outlook Handbook, US Bureau of Labor Statistics, last modified September 1, 2020, https://www.bls.gov /ooh/healthcare/massage-therapists.htm.

30. Edmund S. Higgins, "Is Mental Health Declining in the U.S.?," *Scientific American Mind*, January 1, 2017, https://www.scientificamerican.com /article/is-mental-health-declining-in-the-u-s.

31. Jon Fortenbury, "Fighting Loneliness with Cuddle Parties," *Atlantic*, July 15, 2014, https://www.theatlantic.com/health/archive/2014/07/fighting -loneliness-with-cuddle-parties/373335; Yoni Alkan, "Cuddlist," 2020, https://www.yonialkan.com/cuddlist.

32. "Madelon Guinazzo Co-Founder and Creator of the Certification Program of Cuddlist," December 4, 2017, https://cuddlist.typeform.com/to /csvtzh?afmc=xxxxx.

33. "Cuddle Party," http://www.cuddleparty.com.

34. Kathleen M. Cumiskey and Larissa Hjorth, "'I Wish They Could Have Answered Their Phones': Mobile Communication in Mass Shootings," *Death Studies* 43, no. 7 (December 31, 2018): 414–425, doi:10.1080/ 07481187.2018.1541940; Michelle Drouin, Brandon T. McDaniel, Jessica A. Pater, and Tammy Toscos, "How Parents and Their Children Used Social Media and Technology at the Beginning of the Covid-19 Pandemic and Associations with Anxiety," *Cyberpsychology, Behavior, and Social Networking* 23, no. 11 (November 6, 2020): 727–736, doi:10.1089/ cyber.2020.0284.

35. Kathleen M. Cumiskey and Larissa Hjorth, *Haunting Hands: Mobile Media Practices and Loss* (New York: Oxford University Press, 2017).

36. Simon Mair, "How Will Coronavirus Change the World?," *BBC*, March 31, 2020, https://www.bbc.com/future/article/20200331-covid-19-how-will-the-coronavirus-change-the-world.

37. Martha Henriques, "Will Covid-19 Have a Lasting Impact on the Environment?," *BBC*, March 27, 2020, https://www.bbc.com/future/article/20200326-covid-19-the-impact-of-coronavirus-on-the-environment; Aaron James, *Surfing with Sartre: An Aquatic Inquiry into a Life of Meaning* (New York: Doubleday, 2017).

38. Stevan E. Hobfoll, Charles D. Spielberger, Shlomo Breznitz, Charles Figley, Susan Folkman, Bonnie Lepper-Green, Donald Meichenbaum, et al., "War-Related Stress: Addressing the Stress of War and Other Traumatic Events," *American Psychologist* 46, no. 8 (1991): 848–855, doi:10.1037/0003–066X.46.8.848.

CHAPTER 2

1. Elliott M. Blass and Lisa B. Hoffmeyer, "Sucrose as an Analgesic for Newborn Infants," *Pediatrics* 87, no. 2 (February 1991): 215.

2. Ruth Feldman, Zehava Rosenthal, and Arthur I. Eidelman, "Maternal-Preterm Skin-to-Skin Contact Enhances Child Physiologic Organization and Cognitive Control across the First 10 Years of Life," *Biological Psychiatry* 75, no. 1 (January 2014): 56, doi:10.1016/j.biopsych.2013.08.012; Alex Smith, "Babies Born Dependent on Opioids Need Touch, Not Tech," *NPR*, August 16, 2018, https://www.npr.org/sections/health-shots/2018/08/16/632697780/babies-born-dependent-on-opioids-need-touch-not-tech.

3. Veeral N. Tolia, Stephen W. Patrick, Monica M. Bennett, Karna Murthy, John Sousa, P. Brian Smith, Reese H. Clark, and Alan R. Spitzer, "Increasing Incidence of the Neonatal Abstinence Syndrome in U.S. Neonatal ICUs," *New England Journal of Medicine* 372, no. 22 (May 2015): 2118–2126, doi:10.1056/NEJMsa1500439.

4. Tyler N. A. Winkelman, Nicole Villapiano, Katy B. Kozhimannil, Matthew M. Davis, and Stephen W. Patrick, "Incidence and Costs of Neonatal Abstinence Syndrome among Infants with Medicaid," *Pediatrics* 141,

no. 4 (April 2018): e20173520, doi:10.1542/peds.2017–3520; Shahla M. Jilani, Meghan T. Frey, Dawn Pepin, Tracey Jewell, Melissa Jordan, Angela M. Miller, Meagan Robinson, et al., "Evaluation of State-Mandated Reporting of Neonatal Abstinence Syndrome—Six States, 2013–2017," *Morbidity and Mortality Weekly Report* 68, no. 1 (January 2019): 6–10, doi:dx.doi.org/10.15585/mmwr.mm6801a2.

5. Ian Zuzarte, Premananda Indic, Bruce Barton, David Paydarfar, Francis Bednarek, and Elisabeth Bloch-Salisbury, "Vibrotactile Stimulation: A Non-Pharmacological Intervention for Opioid-Exposed Newborns," *PloS One* 12, no. 4 (April 2017): e0175981, doi:10.1371/journal. pone.0175981; Amanda Hignell, Karen Carlyle, Catherine Bishop, Mary Murphy, Teresa Valenzano, Suzanne Turner, and Michael Sgro, "The Infant Cuddler Study: Evaluating the Effectiveness of Volunteer Cuddling in Infants with Neonatal Abstinence Syndrome," *Paediatrics and Child Health* 25, no. 7 (November 2020): 414–418, doi:10.1093/pch /pxz127.

6. Emis Akbari, Noam Binnoon-Erez, Michelle Rodrigues, Alessandro Ricci, Juliane Schneider, Sheri Madigan, and Jennifer Jenkins, "Kangaroo Mother Care and Infant Biopsychosocial Outcomes in the First Year: A Meta-Analysis," *Early Human Development* 122 (July 2018): 22–31, doi:10.1016/j.earlhumdev.2018.05.004.

7. Rivka Landau, "Affect and Attachment: Kissing, Hugging, and Patting as Attachment Behaviors," *Infant Mental Health Journal* 10, no. 1 (Spring 1989): 59–69, doi:10.1002/1097–0355(198921)10:1<59::AID-IMHJ2280100106>3.0.CO;2–6.

8. Michael Inbar, "Mom's Hug Revives Baby That Was Pronounced Dead," *Today*, September 9, 2010, https://www.today.com/parents/moms-hug -revives-baby-was-pronounced-dead-2D80554298.

9. Naveeen A, "Premature Baby Presumed Dead Comes Backto Life after Mother Cuddles Him: 'Power Of Love,'" *Shared*, November 21, 2019, https://life.shared.com/dead-premature-baby-comes-back-to-life-after -mother-cuddles-him.

10. René A. Spitz, "Hospitalism—an Inquiry into the Genesis of Psychiatric Conditions in Early Childhood," *Psychoanalytic Study of the Child* 1 (1945): 53–74; René A. Spitz, "The Role of Ecological Factors in

Emotional Development in Infancy," *Child Development* 20, no. 3 (September 1949): 145–155, https://doi.org/10.2307/1125870.

11. Harry F. Harlow and Robert R. Zimmermann, "The Development of Affective Responsiveness in Infant Monkeys," *Proceedings of the American Philosophical Society* 102, no. 5 (October 1958): 501–509, https://www.jstor.org/stable/985597.

12. Daksha Trivedi, "Cochrane Review Summary: Massage for Promoting Mental and Physical Health in Typically Developing Infants under the Age of Six Months," *Primary Health Care Research and Development* 16, no. 1 (January 2015): 3–4, doi:10.1017/S1463423614000462.

13. Colin Hesse, Alan C. Mikkelson, and Stephanie Saracco, "Parent-Child Affection and Helicopter Parenting: Exploring the Concept of Excessive Affection," *Western Journal of Communication* 82, no. 4 (August 2017): 457–474, doi:10.1080/10570314.2017.1362705.

14. Inge Bretherton and Kristine A. Munholland, "Internal Working Models in Attachment Relationships: Elaborating a Central Construct in Attachment Theory," in *Handbook of Attachment: Theory, Research, and Clinical Applications*, ed. Jude Cassidy and Philip R. Shaver (New York: Guilford Press, 2008), 102–127.

15. "Women in the Labor Force," Women's Bureau, US Department of Labor, https://www.dol.gov/agencies/wb/data/facts-over-time/women-in-the-labor-force#civilian-labor-force-by-sex.

16. Juliana Menasce Horowitz, "Despite Challenges at Home and Work, Most Working Moms and Dads Say Being Employed Is What's Best for Them," Pew Research Center, https://www.pewresearch.org/fact-tank/2019/09/12/despite-challenges-at-home-and-work-most-working-moms-and-dads-say-being-employed-is-whats-best-for-them.

17. Sara Kettler, "Fred Rogers Took a Stand against Racial Inequality When He Invited a Black Character to Join Him in a Pool," *Biography*, last modified June 24, 2020, https://www.biography.com/news/mister-rogers-officer-clemmons-pool.

18. Maxwell King, "How 'Mister Rogers' Neighborhood' Championed Children with Disabilities," *Guideposts*, October 15, 2018, https://www.guideposts.org/inspiration/inspiring-stories/how-mister-rogers-neighborhood-championed-children-with-disabilities.

19. "Disney Expands Its 'Baby Einstein' Refunds," *CBS News*, October 24, 2009, https://www.cbsnews.com/news/disney-expands-its-baby-einstein-refunds.

20. Sarah Roseberry Lytle, Adrian Garcia-Sierra, and Patricia K. Kuhl, "Two Are Better than One: Infant Language Learning from Video Improves in the Presence of Peers," *Proceedings of the National Academy of Sciences* 115, no. 40 (October 2018): 9859–9866, doi:10.1073/pnas.1611621115.

21. Brandon T. McDaniel and Sarah M. Coyne, "Technology Interference in the Parenting of Young Children: Implications for Mothers' Perceptions of Coparenting," *Social Science Journal* 53, no. 4 (December 2016): 435–443, doi:10.1016/j.soscij.2016.04.010; Genni Newsham, Michelle Drouin, and Brandon T. McDaniel, "Problematic Phone Use, Depression, and Technology Interference among Mothers," *Psychology of Popular Media Culture* 9, no. 2 (December 2018): 117–124, https://doi.org/10.1037/ppm0000220.

22. John Bowlby, *Attachment and Loss, Volume 1: Attachment* (London: Penguin Books, 1978).

23. Nellie Bowles, "The Digital Gap between Rich and Poor Kids Is Not What We Expected," *New York Times*, October 26, 2018, https://www.nytimes.com/2018/10/26/style/digital-divide-screens-schools.html.

24. Paige Leskin, "Twitter CEO Jack Dorsey Uses His iPhone's Screen Time Feature to Limit His Twitter Use to 2 Hours a Day," *Business Insider*, October 25, 2019, https://www.businessinsider.com/jack-dorsey-iphone-screen-time-feature-2-hour-twitter-limit-2019-10.

25. World Health Organization, *Guidelines on Physical Activity, Sedentary Behaviour and Sleep for Children under 5 Years of Age*, 2019, https://apps.who.int/iris/handle/10665/311664.

26. Rebecca Muller, "Evan Spiegel and Miranda Kerr Allow Their Kids 90 Minutes of Screen Time per Week. What's the Right Number for You?," *Thrive Global*, January 2, 2019, https://thriveglobal.com/stories/how-much-weekly-screen-time-limits-kids-children-tips.

27. Olivia Rudugard, "The Tech Moguls Who Invented Social Media Have Banned Their Children from It," *Independent.ie*, November 6, 2018, https://www.independent.ie/life/family/parenting/the-tech-moguls-who

-invented-social-media-have-banned-their-children-from-it-37494367
.html.

28. "Media and Technology Philosophy," Waldorf School of the Peninsula,
https://waldorfpeninsula.org/curriculum/media-technology-philosophy.

29. Parks Associates, *Households with Children: Dominant Tech Purchasers*,
2020, https://www.parksassociates.com/marketfocus/digital-parenting.

30. Sandra L. Calvert, "Children as Consumers: Advertising and Marketing," *Future of Children* 18, no. 1 (Spring 2008): 205–234, doi:10.1353/
foc.0.0001.

31. Brooke Auxier, Monica Anderson, Andrew Perrin, and Erica Turner, "Parenting Children in the Age of Screens," Pew Research Center, July 28, 2020,
https://www.pewresearch.org/internet/2020/07/28/parenting-children
-in-the-age-of-screens.

32. "The Common Sense Census: Media Use by Tweens + Teens," Common
Sense Media, November 3, 2015, https://www.commonsensemedia.org
/the-common-sense-census-media-use-by-tweens-and-teens-infographic.

33. Bowles, "The Digital Gap."

34. Gregg Weiss, "Wow, i spend more time on Twitter than Jack Dorsey! @
BuzzFeedBen: How much time do you spend on Twitter per day? @jack:
I set the screen time limit to 2 hours . . . ," Twitter, October 24, 2019,
https://twitter.com/greggweiss/status/1187466588506558464.

35. "The Common Sense Census."

36. "Welcome to the Harvard Study of Adult Development," Harvard Second
Generation Study, Harvard Medical School, https://www.adultdevelop
mentstudy.org.

37. Diana C. Reep and Faye H. Dambrot, "Effects of Frequent Television View-
ing on Stereotypes: 'Drip, Drip' or 'Drench'?," *Journalism Quarterly* 66, no.
3 (September 1989): 542–550, doi:10.1177/107769908906600302.

CHAPTER 3

1. Bratislav Stojiljkovic, Dragoljub A. Cucic, and Zoran Pajic, "Nikola Tesla
and Samuel Clemens: The Friendship between Two Luminaries of the
Gilded Age," *Mark Twain Journal* 52, no. 2 (November 2014): 25–62.

2. Lydia Denworth, *Friendship: The Evolution, Biology, and Extraordinary Power of Life's Fundamental Bond* (New York: W. W. Norton and Company, 2020).

3. Robert Plomin, J. C. DeFries, and John C. Loehlin, "Genotype-Environment Interaction and Correlation in the Analysis of Human Behavior," *Psychological Bulletin* 84, no. 2 (1977): 309–322, doi:10.1037/0033-2909.84.2.309.

4. "Girl Scouts," Girl Scouts of the United States of America, https://www.girlscouts.org.

5. Miller McPherson, Lynn Smith-Lovin, and James M. Cook, "Birds of a Feather: Homophily in Social Networks," *Annual Review of Sociology* 27 (August 2001): 415–444, doi:10.1146/annurev.soc.27.1.415.

6. Erik Erikson, *Childhood and Society* (New York: W. W. Norton and Company, 1950).

7. "Erik Erikson, 91, Psychoanalyst Who Reshaped Views of Human Growth, Dies," *New York Times*, May 13, 1994, https://www.nytimes.com/1994/05/13/obituaries/erik-erikson-91-psychoanalyst-who-reshaped-views-of-human-growth-dies.html?pagewanted=all.

8. Melanie Curtin, "This 75-Year Harvard Study Found the 1 Secret to Leading a Fulfilling Life," *Inc.*, February 27, 2017, https://www.inc.com/melanie-curtin/want-a-life-of-fulfillment-a-75-year-harvard-study-says-to-prioritize-this-one-t.html.

9. John M. Reisman, "An Indirect Measure of the Value of Friendship for Aging Men," *Journal of Gerontology* 43, no. 4 (July 1988): 109–110, https://doi.org/10.1093/geronj/43.4.P109.

10. Denworth, *Friendship*.

11. Julianne Holt-Lunstad, Timothy B. Smith, and J. Bradley Layton, "Social Relationships and Mortality Risk: A Meta-Analytic Review," *PLoS Medicine* 7, no. 7 (July 2010): 1–20, doi:10.1371/journal.pmed.1000316.

12. Stephanie Buck, "1980s Teens Spent Thousands Flirting with Strangers on the Phone," *Timeline*, February 6, 2017, https://timeline.com/party-lines-teens-a1be20c45686.

13. "PenPal World: Where People Become Friends," Penpal World, Inc., http://www.penpalworld.com; "PenPal Schools," PenPal Schools, https://www.penpalschools.com.

14. John Mayer, Twitter post, October 2017, 3:51 p.m., https://twitter.com/johnmayer/status/915665811892609024.

15. Lindsay Dodgson, "4 People on the Average Flight Could Meet the Love of Their Life in the Air, According to a Survey," *Insider*, August 30, 2018, https://www.insider.com/one-in-50-passengers-meet-the-love-of-their-life-on-a-plane-2018-8.

16. "Imagine Better Future with Your DNA," MyGenomeBox, https://www.mygenomebox.com.

17. "Amazon Prime Air," Amazon.com, https://www.amazon.com/Amazon-Prime-Air/b?ie=UTF8&node=8037720011; Annie Palmer, "Amazon Wins FAA Approval for Prime Air Drone Delivery Fleet," *CNBC*, August 31, 2020, https://www.cnbc.com/2020/08/31/amazon-prime-now-drone-delivery-fleet-gets-faa-approval.html.

18. Nicholas Epley and Juliana Schroeder, "Mistakenly Seeking Solitude," *Journal of Experimental Psychology: General* 143, no. 5 (October 2014): 1980, doi:10.1037/a0037323.

19. Stanley Milgram and John Sabini, "On Maintaining Social Norms: A Field Experiment in the Subway," in *Advances in Environmental Psychology, Vol. 1: The Urban Environment*, ed. Jerome E. Singer and Stuart Valins (Hillsdale, NJ: Erlbaum, 1978), 31–40.

20. Susanne Shultz, Christopher Opie, and Quentin D. Atkinson, "Stepwise Evolution of Stable Sociality in Primates," *Nature* 479 (2011): 219–222.

21. John M. Zelenski, Deanna C. Whelan, Logan J. Nealis, Christina M. Besner, Maya S. Santoro, and Jessica E. Wynn, "Personality and Affective Forecasting: Trait Introverts Underpredict the Hedonic Benefits of Acting Extraverted," *Journal of Personality and Social Psychology* 104, no. 6 (April 2013): 1092–1108, doi:10.1037/a0032281.

22. Holt-Lunstad, Smith, and Layton, "Social Relationships and Mortality Risk."

23. Ingibjorg E. Thorisdottir, Rannveig Sigurvinsdottir, Alfgeir L. Kristjansson, John P. Allegrante, Christa L. Lilly, and Inga Dora Sigfusdottir,

"Longitudinal Association between Social Media Use and Psychological Distress among Adolescents," *Preventive Medicine* 1410 (December 2020): 106270, doi:10.1016/j.ypmed.2020.106270.

24. Brian A. Primack, Ariel Shensa, Jaime E. Sidani, Erin O. Whaite, Liu Yi Lin, Daniel Rosen, Jason B. Colditz, et al., "Social Media Use and Perceived Social Isolation among Young Adults in the US," *American Journal of Preventive Medicine* 53, no. 1 (March 2017): 1–8, doi:10.1016/j.amepre.2017.01.010.

25. Melissa G. Hunt, Rachel Marx, Courtney Lipson, and Jordyn Young, "No More FOMO: Limiting Social Media Decreases Loneliness and Depression," *Journal of Social and Clinical Psychology* 37, no. 10 (December 2018): 751–768, doi:10.1521/jscp.2018.37.10.751.

26. Jonathan Haidt and Jean Twenge, "Social Media Use and Mental Health: A Review" (unpublished manuscript, New York University, 2019).

27. Ethan Kross, Philippe Verduyn, Gal Sheppes, Cory K. Costello, John Jonides, and Oscar Ybarra, "Social Media and Well-Being: Pitfalls, Progress, and Next Steps," *Trends in Cognitive Sciences* 25, no. 1 (November 2020): 55–66, doi:10.1016/j.tics.2020.10.005.

28. Gary M. Cooney, Kerry Dwan, Carolyn A. Greig, Debbie A. Lawlor, Jane Rimer, Fiona R. Waugh, Marion McMurdo, and Gillian E. Mead, "Exercise for Depression," *Cochrane Database of Systematic Reviews* 9 (September 2013), CD004366, doi:10.1002/14651858.CD004366.pub6.

29. Lauren J. N. Brent, Steve W. C. Chang, Jean-François Gariépy, and Michael L. Platt, "The Neuroethology of Friendship," *Annals of the New York Academy of Sciences* 1316, no. 1 (December 2013): 1–17, doi:10.1111/nyas.12315.

30. Robin I. M. Dunbar, "The Social Brain Hypothesis," *Evolutionary Anthropology* 6, no. 5 (December 1998): 178–190, https://doi.org/10.1002/(SICI)1520–6505(1998)6:5<178::AID-EVAN5>3.0.CO;2–8.

31. Suzanne N. Haber, "Neuroanatomy of Reward: A View from the Ventral Striatum," in *Neurobiology of Sensation and Reward*, ed. Jay A. Gottfried and Jay A. Gottfried (Boca Raton, FL: CRC Press, 2011), 235–261.

32. Mattie Tops, Sander L. Koole, Hans Ijzerman, and Femke T. A. Busman-Pijlman, "Why Social Attachment and Oxytocin Protect against Addiction

and Stress: Insights from the Dynamics between Ventral and Dorsal Corticostriatal Systems," *Pharmacology, Biochemistry and Behavior* 119 (April 2014): 39–48, doi:10.1016/j.pbb.2013.07.015.

33.   Ed Diener, Martin E. P. Seligman, Hyewon Choi, and Shigehiro Oishi, "Happiest People Revisited," *Perspectives on Psychological Science* 13, no. 2 (March 2018): 176–184, doi:10.1177/1745691617697077.

34.   Bianca DiJulio, Liz Hamel, Cailey Muñana, and Mollyann Brodie, "Loneliness and Social Isolation in the United States, the United Kingdom, and Japan: An International Survey," Kaiser Family Foundation, August 30, 2018, https://www.kff.org/other/report/loneliness-and-social-isolation-in -the-united-states-the-united-kingdom-and-japan-an-international-survey.

35.   Emily A. Kuhl, "Quantifying the Cost of Depression," Center for Workplace Mental Health, American Psychiatric Association, http://www .workplacementalhealth.org/Mental-Health-Topics/Depression /Quantifying-the-Cost-of-Depression.

36.   "The Cost of Loneliness Project," Cost of Loneliness Project, https://www .thecostofloneliness.org; Julianne Holt-Lunstad, "So Lonely I Could Die," Parenting, Family, and Relationships, American Psychological Association, August 5, 2017, https://www.apa.org/news/press/releases/2017/08 /lonely-die.

37.   Beth Gillette, "I Tried Bumble BFF for 30 Days—Here's What Happened," Every Girl, October 4, 2018, https://theeverygirl.com/i-tried-bumble -bff-for-30-days-heres-what-happened.

38.   "Family Romance," Family Romance, http://family-romance.com/; Roc Morin, "How to Hire Fake Friends and Family," *Atlantic*, November 7, 2017, https://www.theatlantic.com/family/archive/2017/11/paying-for-fake -friends-and-family/545060.

39.   Kelly Campbell, Nicole Holderness, and Matt Riggs, "Friendship Chemistry: An Examination of Underlying Factors," *Social Science Journal* 52, no. 2 (December 2019): 239–247. doi:10.1016/j.soscij.2015.01.005.

40.   Jean-Luc Bouchard, "I Paid $47 an Hour for Someone to Be My Friend," *Vox*, last modified July 1, 2019, https://www.vox.com/the-high light/2019/6/24/18701140/rent-a-friend-friendship-loneliness-platonic -relationships.

CHAPTER 4

1. Phillip Steadman, "Samuel Bentham's Panopticon," *Journal of Bentham Studies* 14, no. 1 (January 2012): 1–30, doi:10.14324/111.2045–757X.044.

2. Alberto Romele, Francesco Gallino, Camilla Emmenegger, and Daniele Gorgone, "Panopticism Is Not Enough: Social Media as Technologies of Voluntary Servitude," *Surveillance and Society* 15, no. 2 (February 2017): 204–221, https://doi.org/10.24908/ss.v15i2.6021.

3. David Lyon, "Surveillance, Snowden, and Big Data: Capacities, Consequences, Critique," *Big Data and Society* (July 2014): 1–13, https://doi.org/10.1177/2053951714541861; Nicholas Confessore, "Cambridge Analytica and Facebook: The Scandal and the Fallout So Far," *New York Times*, April 4, 2018, https://www.nytimes.com/2018/04/04/us/politics/cambridge-analytica-scandal-fallout.html; Paolo Zialcita, "Facebook Pays $643,000 Fine for Role in Cambridge Analytica Scandal," *NPR*, October 30, 2019, https://www.npr.org/2019/10/30/774749376/facebook-pays-643-000-fine-for-role-in-cambridge-analytica-scandal.

4. Ben Wolford, "What Is GDPR, the EU's New Data Protection Law?," Horizon 2020 Framework Programme of the European Union, https://gdpr.eu/what-is-gdpr.

5. Romele et al., "Panopticism Is Not Enough."

6. Stanely Cohen, *States of Denial: Knowing about Atrocities and Suffering* (Cambridge, UK: Polity Press, 2001).

7. Jacob Poushter, Caldwell Bishop, and Hanyu Chwe, "Social Media Use Continues to Rise in Developing Countries, but Plateaus across Developed Ones," Pew Research Center, June 19, 2018, https://www.pewresearch.org/global/2018/06/19/3-social-network-adoption-varies-widely-by-country.

8. Michael Winnick, "Putting a Finger on Our Phone Obsession," *dscout* (blog), June 16, 2016, https://blog.dscout.com/mobile-touches.

9. Brandon T. McDaniel and Sarah Coyne, "'Technoference': The Interference of Technology in Couple Relationships and Implications for Women's Personal and Relational Wellbeing," *Psychology of Popular Media Culture* 5, no. 1 (December 2014): 85–98, doi:10.1037/ppm0000065; James A. Roberts and Meredith E. David, "My Life Has Become a Major Distraction from My Cell Phone: Partner Phubbing and Relationship Satisfaction

among Romantic Partners," *Computers in Human Behavior* 54 (January 2016): 134–141, doi:10.1016/j.chb.2015.07.058.

10. Daniel Halpern and James E. Katz, "Texting's Consequences for Romantic Relationships: A Cross-Lagged Analysis Highlights Its Risks," *Computers in Human Behavior* 71 (June 2017): 386–394, https://doi.org/10.1016/j .chb.2017.01.051; Hanna Krasnova, Olga Abramova, Isabelle Notter, and Annika Baumann, "Why Phubbing Is Toxic for Your Relationship: Understanding the Role of Smartphone Jealousy among 'Generation Y' Users" (paper presented at the twenty-fourth European Conference on Information Systems, Istanbul, June 2016); Xingchao Wang, Xiaochun Xie, Yuhui Wang, and Pengcheng Wang, "Partner Phubbing and Depression among Married Chinese Adults: The Roles of Relationship Satisfaction and Relationship Length," *Personality and Individual Differences* 110 (May 2017): 12–17, doi:10.1016/j.paid.2017.01.014.

11. Brandon T. McDaniel and Michelle Drouin, "Daily Technology Interruptions and Emotional and Relational Well-Being," *Computers in Human Behavior* 99 (October 2019): 1–8, doi:10.1016/j.chb.2019.04.027.

12. Norman K. Denzin, *Symbolic Interactionism and Cultural Studies: The Politics of Interpretation* (Cambridge, MA: Blackwell, 2008).

13. Mariek Vanden Abeele, Marjolijn L. Antheunis, and Alexander P. Schouten, "The Effect of Mobile Messaging during a Conversation on Impression Formation and Interaction Quality," *Computers in Human Behavior* 62 (September 2016): 562–569.

14. John W. Thibault and Harold H. Kelley, *The Social Psychology of Groups* (New York: Wiley, 1959).

15. *The Social Dilemma*, Exposure Labs, https://www.thesocialdilemma.com.

16. Jean M. Twenge, Thomas E. Joiner, Megan L. Rogers, and Gabrielle N. Martin, "Increases in Depressive Symptoms, Suicide-Related Outcomes, and Suicide Rates among U.S. Adolescents after 2010 and Links to Increased New Media Screen-Time," *Clinical Psychological Science* 6, no. 1 (November 2017): 3–17; Jean M. Twenge, Gabrielle N. Martin, and W. Keith Campbell, "Decreases in Psychological Well-Being among American Adolescents after 2012 and Links to Screen-Time during the Rise of Smartphone Technology," *Emotion* 18, no. 6 (January 2018): 765–780, doi:10.1037/emo0000403.

17. Rose Maghsoudi, Jennifer Shapka, and Pamela Wisniewski, "Examining How Online Risk Exposure and Online Social Capital Influence Adolescent Psychological Stress," *Computers in Human Behavior* 113 (December 2020): 106488, doi:10.1016/j.chb.2020.106488; Melissa G. Hunt, Jordyn Young, Rachel Marx, and Courtney Lipson, "No More FOMO: Limiting Social Media Decreases Loneliness and Depression," *Journal of Social and Clinical Psychology* 37, no. 10 (December 2018): 751–768, https://doi.org/10.1521/jscp.2018.37.10.751; Madeleine J. George, "Concurrent and Subsequent Associations between Daily Digital Technology Use and High-Risk Adolescents' Mental Health Symptoms," *Child Development* 89, no. 1 (May 2017): 78–88, doi:10.1111/cdev.12819.

18. Amy Orben and Andrew K. Przybylski, "The Association between Adolescent Well-Being and Digital Technology Use," *Nature Human Behaviour* 3, no. 2 (February 2019): 173–182, doi:10.1038/s41562-018-0506-1; Rebecca Nowland, Elizabeth A. Necka, and John T. Cacioppo, "Loneliness and Social Internet Use: Pathways to Reconnection in a Digital World?," *Perspectives on Psychological Science* 13, no 1. (September 2017): 70–87, doi:10.1177/1745691617713052.

19. Lisa F. Berkman and S. Leonard Syme, "Social Networks, Host Resistance, and Mortality: A Nine-Year Follow-up of Alameda County Residents," *American Journal of Epidemiology* 109, no. 2 (February 1979): 186–204, https://doi.org/10.1093/oxfordjournals.aje.a112674.

20. Sheldon Cohen, William J. Doyle, David P. Skoner, Bruce S. Rabin, and Jack M. Gwaltney Jr., "Social Ties and Susceptibility to the Common Cold," *Journal of the American Medical Association* 277, no. 24 (June 1997): 1940–1944.

21. Sheldon Cohen, "Social Network Index," http://www.midss.org/sites/default/files/social_network_index.pdf.

22. Nicole B. Ellison, Charles Steinfield, and Cliff Lampe, "The Benefits of Facebook 'Friends': Exploring the Relationship between College Students' Use of Online Social Networks and Social Capital," *Journal of Computer-Mediated Communication* 12, no. 3 (July 2007): 1143–1168, https://doi.org/10.1111/j.1083-6101.2007.00367.x; Nan Lin, *Social Capital: A Theory of Social Structure and Action* (New York: Cambridge University Press, 2001).

23. Rose Maghsoudi, Jennifer Shapka, and Pamela Wisniewski, "Examining How Online Risk Exposure and Online Social Capital Influence Adolescent Psychological Stress," *Computers in Human Behavior* 113 (December 2020): 106488, doi:10.1016/j.chb.2020.106488.

24. Thomas Ashby Wills, "Social Support and Interpersonal Relationships," *Review of Personality and Social Psychology* 12 (1991): 265–289; Sheldon Cohen and S. Leonard Syme, *Social Support and Health* (New York: Academic, 1985); James S. House, "Social Support and Social Structure," *Sociological Forum* 2, no. 1 (1987): 135–146, http://dx.doi.org/10.1007/BF01107897.

25. Sheldon Cohen and Garth McKay, "Social Support, Stress, and the Buffering Hypothesis: A Theoretical Analysis," in *Handbook of Psychology and Health,* ed. Shelley E. Taylor, Jerome Singer, and Andrew Baum (Abingdon, UK: Routledge, 1984), 253–268; Stephanie L. Brown, Randolph M. Nesse, Amiram D. Vinokur, and Dylan M. Smith, "Providing Social Support May Be More Beneficial Than Receiving It: Results from a Prospective Study of Mortality," *Psychological Science* 14, no. 4 (July 2003): 320–327, doi:10.1111/1467-9280.14461; Ralf Schwarzer and Anja Leppin, "Social Support and Health: A Meta-Analysis," *Psychology and Health* 3, no. 1 (August 1988): 1–15, doi:10.1080/08870448908400361; Faith Ozbay, Douglas C. Johnson, Eleni Dimoulas, C. A. Morgan III, Dennis Charney, and Steven Southwick, "Social Support and Resilience to Stress: From Neurobiology to Clinical Practice," *Psychiatry* 4, no. 5 (May 2007): 35–40; James S. House, Karl R. Landis, and Debra Umberson, "Social Relationships and Health," *Science* 241, no. 4865 (July 1988): 540–545, doi:10.1126/science.3399889; Candyce H. Kroenke, Laura D. Kubzansky, Eva S. Schernhammer, Michelle D. Holmes, and Ichiro Kawachi, "Social Networks, Social Support, and Survival after Breast Cancer Diagnosis," *Journal of Clinical Oncology* 24, no. 7 (March 2006), 1105–1111, doi:10.1200/JCO.2005.04.2846.

26. Catherine A. Heany and Barbara A. Israel, "Social Networks and Social Support," in *Health Behavior and Health Education: Theory, Research and Practice*, 3rd ed., ed. Karen Glanz, Barbara K. Rimer, and K. Viswanath (San Francisco: Jossey-Bass Publishing, 2002), 189–210.

27. Mufan Luo and Jeffrey T. Hancock, "Self-Disclosure and Social Media: Motivations, Mechanisms and Psychological Well-Being," *Current*

Opinion in Psychology 31 (February 2020): 110–115. doi:10.1016/j.copsyc.2019.08.019.

28. Trevor Haynes, "Dopamine, Smartphones and You: A Battle for Your Time," Graduate School of Arts and Sciences, Harvard University, Science in the News, http://sitn.hms.harvard.edu/flash/2018/dopamine-smartphones-battle-time.

29. Irving Biederman and Edward A. Vessel, "Perceptual Pleasure and the Brain," *American Scientist* 94, no. 3 (May 2006): 247, doi:10.1511/2006.3.247.

30. Marie Kondo, *Life-Changing Magic of Tidying Up* (Berkeley, CA: Ten Speed Press, October 2014).

31. Julie Tseng and Jordan Poppenk, "Brain Meta-State Transitions Demarcate Thoughts across Task Contexts Exposing the Mental Noise of Trait Neuroticism," *Nature Communications* 11 (July 2020): 3480, https://doi.org/10.1038/s41467-020-17255-9.

32. Susan Nolen-Hoeksema, "The Role of Rumination in Depressive Disorders and Mixed Anxiety/Depressive Symptoms," *Journal of Abnormal Psychology* 109, no. 3 (July 2000): 504–511, doi:10.1037/0021–843X.109.3.504.

33. Alexander M. Penney, Victoria C. Miedema, and Dwight Mazmanian, "Intelligence and Emotional Disorders: Is the Worrying and Ruminating Mind a More Intelligent Mind?," *Personality and Individual Differences* 74 (February 2015): 90–93, doi:10.1016/j.paid.2014.10.005.

34. Belgin Ünal and Miri Besken, "Blessedly Forgetful and Blissfully Unaware: A Positivity Bias in Memory for (Re)constructions of Imagined Past and Future Events," *Memory* 28, no. 7 (August 2020): 888–899, doi:10.1080/09658211.2020.1789169.

35. Margaret W. Matlin and David J. Stang, *The Pollyanna Principle: Selectivity in Language, Memory, and Thought* (Cambridge, MA: Schenkman, 1978).

36. Margaret W. Matlin and Valerie J. Gawron, "Individual Differences in Pollyannaism," *Journal of Personality Assessment* 43, no. 4 (June 2010): 411–412, doi:10.1207/s15327752jpa4304_14.

37. Jesse Fox and Bree McEwan, "Distinguishing Technologies for Social Interaction: The Perceived Social Affordances of Communication Channels Scale," *Communication Monographs* 84, no. 3 (June 2017): 298–318, doi:10.1080/03637751.2017.1332418.

38. Bridget Pujals and Manik Singh, "Swipe Up! Discover Vanish Mode for Messenger and Instagram," *Messenger News*, Facebook, November 12, 2020, https://messengernews.fb.com/2020/11/12/swipe-up-discover -vanish-mode-for-messenger-and-instagram.

39. Albert Bandura, *Social Learning Theory* (Saddle River, NJ: Prentice Hall, 1977).

40. Abraham Maslow, "A Theory of Human Motivation," *Psychological Review* 50, no. 4 (1943): 370–396, doi:10.1037/h0054346.

41. Charles R. Berger and James J. Bradac, *Language and Social Knowledge: Uncertainty in Interpersonal Relations* (London: Edward Arnold, 1982); Charles R. Berger and Richard J. Calabrese, "Some Explorations in Initial Interaction and Beyond: Toward a Developmental Theory of Interpersonal Communication," *Human Communication Research* 1, no. 2 (January 1975): 99–112, doi:10.1111/j.1468–2958.1975.tb00258.x.

42. Leah E. LeFebvre and Xiaoti Fan, "Ghosted?: Navigating Strategies for Reducing Uncertainty and Implications Surrounding Ambiguous Loss," *Personal Relationships* 27, no. 2 (June 2020): 433–459, doi:10.1111 /pere.12322.

43. Arie W. Kruglanski, "Motivations for Judging and Knowing: Implications for Causal Attribution," in *Handbook of Motivation and Cognition: Foundations of Social Behavior, Vol. 2*, ed. E. Tory Higgins and Richard M. Sorrentino (New York: Guilford Press, 1990), 333–368.

44. Megan Johnson, "YJ Tried It: Dopamine Fasting Helped Me Appreciate the Present," *Yoga Journal*, September 13, 2020, https://www.yogajournal .com/lifestyle/dopamine-fasting-to-appreciate-the-present; Nellie Bowles, "How to Feel Nothing Now, in Order to Feel More Later," *New York Times*, November 7, 2019, https://www.nytimes.com/2019/11/07/style /dopamine-fasting.html.

CHAPTER 5

1. Emily A. Vogels, "10 Facts about American and Online Dating," Pew Research Center, February 6, 2020, https://www.pewresearch.org/fact -tank/2020/02/06/10-facts-about-americans-and-online-dating.

2. Ángel Castro and Juan Ramón Barrada, "Dating Apps and Their Sociodemographic and Psychosocial Correlates: A Systematic Review,"

International Journal of Environmental Research and Public Health 17, no. 18 (September 2020): 6500, https://doi.org/10.3390/ijerph17186500.

3. Monica Anderson, Emily A. Vogels, and Erica Turner, "The Virtues and Downsides of Online Dating," Pew Research Center, February 6, 2020, https://www.pewresearch.org/internet/2020/02/06/the-virtues-and-downsides-of-online-dating.

4. Donn E. Byrne, *The Attraction Paradigm* (New York: Academic, 1971).

5. Yoram Weiss and Robert Willis, "Match Quality, New Information, and Marital Dissolution," *Journal of Labor Economics* 15, no. 1 (1997): 293–329, https://EconPapers.repec.org/RePEc:ucp:jlabec:v:15:y:1997:i:1:p:s293-329; Robert D. Mare, "Five Decades of Educational Assortative Mating," *American Sociological Review* 56, no. 1 (February 1991): 15–32, doi:10.2307/2095670.

6. Gian C. Gonzaga, Steve Carter, and J. Galen Buckwalter, "Assortative Mating, Convergence, and Satisfaction in Married Couples," *Personal Relationships* 17, no. 4 (November 2010): 634–644, doi:10.1111/j.1475–6811.2010.01309.x.

7. Malcom Brynin, Siometta Longhi, and Álvaro Martínez Pérez, "The Social Significance of Homogamy," ISER Working Papers 2008–32, Institute for Social and Economic Research, Essex, UK.

8. Christine R. Schwartz and Robert D. Mare, "Trends in Educational Assortative Marriage from 1940 to 2003," *Demography* 42 (November 2005): 621–646.

9. Brynin, Longhi, and Martínez Pérez, "The Social Significance of Homogamy."

10. Huiping Zhang, Petula S. Y. Ho, and Paul S. F. Yip, "Does Similarity Breed Marital and Sexual Satisfaction?," *Journal of Sex Research* 49, no. 6 (September 2012): 583–593, doi:10.1080/00224499.2011.574240.

11. Yue Qian, "Gender Asymmetry in Educational and Income Assortative Marriage," *Journal of Marriage and Family* 79, no. 2 (September 2016): 318–336, doi:10.1111/jomf.12372.

12. Marianne Bertrand, Emir Kamenica, and Jessica Pan, "Gender Identity and Relative Income within Households," *Quarterly Journal of Economics* 130, no. 2 (May 2015): 571–614, https://doi.org/10.1093/qje/qjv001.

13. Lingshan Zhang, Anthony J. Lee, Lisa M. DeBruine, and Benedict C. Jones, "Are Sex Differences in Preferences for Physical Attractiveness and Good Earning Capacity in Potential Mates Smaller in Countries with Greater Gender Equality?," *Evolutionary Psychology* 17, no. 2 (April 2019): 31146580, doi:10.1177/1474704919852921.

14. David M. Buss and David P. Schmitt, "Sexual Strategies Theory: An Evolutionary Perspective on Human Mating," *Psychological Review* 100, no. 2 (1993): 204–232, https://doi.org/10.1037/0033-295X.100.2.204; Norman P. Li, J. Michael Bailey, Douglas T. Kenrick, and Joan A. W. Linsenmeier, "The Necessities and Luxuries of Mate Preferences: Testing the Tradeoffs," *Journal of Personality and Social Psychology* 82, no. 6 (2002): 947–955.

15. David M. Buss and Todd K. Shackelford, "Attractive Women Want It All: Good Genes, Economic Investment, Parenting Proclivities, and Emotional Commitment," *Evolutionary Psychology* 6, no. 1 (January 2008): 134, doi:10.1177/147470490800600116.

16. David Ong, "Education and Income Attraction: An Online Dating Field Experiment," *Applied Economics* 48, no. 19 (November 2015): 1816–1830, doi:10.1080/00036846.2015.1109039.

17. Brecht Neyt, Sarah Vandenbulcke, and Stijn Baert, "Are Men Intimidated by Highly Educated Women? Undercover on Tinder," *Economics of Education Review* 73 (December 2019): 101914, https://doi.org/10.1016/j.econedurev.2019.101914.

18. Heather Tonnessen, ed., "Astronaut Jonny Kim," NASA, last modified June 2, 2020, https://www.nasa.gov/astronauts/biographies/jonny-kim/biography.

19. Neyt, Vandenbulcke, and Baert, "Are Men Intimidated by Highly Educated Women?," 10.

20. Personal correspondence with Ella, November 24, 2020.

21. Barry Schwartz, *The Paradox of Choice: Why More Is Less* (New York: Harper Perennial, 2004).

22. Sheena S. Iyengar and Mark R. Lepper, "When Choice Is Demotivating: Can One Desire Too Much of a Good Thing?," *Journal of Personality and Social Psychology* 79, no. 6 (January 2001): 995–1006, doi:10.1037/0022-3514.79.6.995.

23. Nancy H. Brinson and Kathrynn R. Pounders, "Match Me If You Can: Online Dating and the Paradox of Choice" (paper presented at the AMA Winter Educators' Conference, Austin, TX, January 2019); Pai-Lu Wu and Wen-Bin Chiou, "More Options Lead to More Searching and Worse Choices in Finding Partners for Romantic Relationships Online: An Experimental Study," *CyberPsychology and Behavior* 12, no. 3 (June 2009): 315–318, doi:10.1089/cpb.2008.0182; Ian Kwok and Annie B. Wescott, "Cyberintimacy: A Scoping Review of Technology-Mediated Romance in the Digital Age," *Cyberpsychology, Behavior, and Social Networking* 23, no. 10 (October 2020): 657–666, http://doi.org/10.1089/cyber.2019.0764.

24. Mu-Li Yang and Wen-Bin Chiou, "Looking Online for the Best Romantic Partner Reduces Decision Quality: The Moderating Role of Choice-Making Strategies," *Cyberpsychology, Behavior, and Social Networking* 13, no. 2 (April 2010): 207–210, doi:10.1089/cyber.2009.0208.

25. Leo Tolstoy, *Anna Karenina* (New York: P. F. Collier and Son, 1917), https://www.bartleby.com/317/1/704.html

26. Daryl J. Bem, "Feeling the Future: Experimental Evidence for Anomalous Retroactive Influences on Cognition and Affect," *Journal of Personality and Social Psychology* 100, no. 3 (March 2011): 407–425, doi:10.1037/a0021524.

27. Charles M. Judd, and Bertram Gawronski, "Editorial Comment," *Journal of Personality and Social Psychology* 100, no. 3 (March 2011): 406, doi:10.1037/0022789.

28. Daryl Bem, Patrizio E. Tressoldi, Thomas Rabeyron, and Michael Duggan, "Feeling the Future: A Meta-Analysis of 90 Experiments on the Anomalous Anticipation of Random Future Events," *F1000Research* 4 (October 2015): 1188, doi:10.12688/f1000research.7177.1.

29. Etzel Cardeña, "The Experimental Evidence for Parapsychological Phenomena: A Review," *American Psychologist* 73, no. 5 (July 2018): 663–677, doi:10.1037/amp0000236.

30. Daryl Bem, "Self-Perception Theory," in *Advances in Experimental Social Psychology*, ed. Leonard Berkowitz (New York: Academic Press, 1972), 1–62.

31. Lauri Nummenmaa, Riitta Hari, Jari K. Hietanen, and Enrico Glerean, "Maps of Subjective Feelings," *Proceedings of the National Academy of*

Sciences of the United States of America 115, no. 37 (September 2018): 9198–9203, https://doi.org/10.1073/pnas.1807390115; Sofia Volynets, Enrico Glerean, Jari K. Hietanen, Riitta Hari, and Lauri Nummenmaa, "Bodily Maps of Emotions Are Culturally Universal," *Emotion* 20, no. 7 (October 2020): 1127–1136, doi:10.1037/emo0000624.supp.

32. Sofia Volynets et al., "Bodily Maps of Emotions."

33. Walter Mischel, *The Marshmallow Test: Mastering Self-Control* (New York: Little, Brown and Company, 2014).

34. Walter Mischel, *Personality and Assessment* (New York: Wiley, 1968).

35. Anderson, Vogels, and Turner, "The Virtues and Downsides of Online Dating."

36. Nadav Klein and Ed O'Brien, "People Use Less Information than They Think to Make Up Their Minds," *Proceedings of the National Academy of Sciences of the United States of America* 115, no. 52 (December 2018): 13222–13227, doi:10.1073/pnas.1805327115.

37. Stephanie Ortigue, Francesco Bianchi-Demicheli, Nisa Patel, Chris Frum, and James W. Lewis, "Neuroimaging of Love: fMRI Meta-analysis Evidence toward New Perspectives in Sexual Medicine," *Journal of Sexual Medicine* 7, no. 11 (August 2010): 3541–3552, doi:10.1111/j.1743 –6109.2010.01999.x.

38. Anderson, Vogels, and Turner, "The Virtues and Downsides of Online Dating."

39. United States Census Bureau, *America's Families and Living Arrangements: 2018*, November 2018, https://www.census.gov/data/tables/2018/demo /families/cps-2018.html.

40. A. W. Geiger and Gretchen Livingston, "8 Facts about Love and Marriage in America," Pew Research Center, February 13, 2019, https://www .pewresearch.org/fact-tank/2019/02/13/8-facts-about-love-and-marriage.

41. UN Women, *Progress of the World's Women 2019–2020: Families in a Changing World*, 2020, https://www.unwomen.org/en/digital-library /progress-of-the-worlds-women?y=2019&y=2019; Geiger and Livingston, "8 Facts about Love and Marriage."

42. Harold H. Kelley and John W. Thibaut, *Interpersonal Relations: A Theory of Interdependence* (New York: Wiley, 1978).

43. Caryl E. Rusbult, "Commitment and Satisfaction in Romantic Associations: A Test of the Investment Model," *Journal of Experimental Social Psychology* 16, no. 2 (March 1980): 172–186, doi:10.1016/0022-1031(80)90007-4.

44. Michelle Drouin, Daniel A. Miller, and Jayson L. Dibble, "Facebook Or Memory—Which Is the Real Threat to Your Relationship?," *Cyberpsychology, Behavior, & Social Networking* 18, no. 10 (October 2015): 561–566, doi:10.1089/cyber.2015.0259

CHAPTER 6

1. Caryl E. Rusbult, Christopher R. Agnew, and Ximena B. Arriaga, "The Investment Model of Commitment Processes," in *Handbook of Theories of Social Psychology*, ed. Paul A. M. Van Lange, Arie W. Kruglanski, and E. Tony Higgins (Thousand Oaks, CA: Sage Publications, 2012), 218–231.

2. Lindsay T. Labrecque and Mark A. Whisman, "Attitudes toward and Prevalence of Extramarital Sex and Descriptions of Extramarital Partners in the 21st Century," *Journal of Family Psychology* 31, no. 7 (May 2017): 952–957, doi:10.1037/fam0000280.

3. Michelle Drouin, "Sexting," *Encyclopedia Brittanica*, December 5, 2018, https://www.britannica.com/topic/sexting.

4. Adam M. Galovan, Michelle Drouin, and Brandon T. McDaniel, "Sexting Profiles in the United States and Canada," *Computers in Human Behavior* 79, no. C (February 2018): 19–29, doi:10.1016/j.chb.2017.10.017.

5. Ashley E. Thompson and Lucia F. O'Sullivan, "Drawing the Line: The Development of a Comprehensive Assessment of Infidelity Judgments," *Journal of Sex Research* 53, no. 8 (November 2015): 910–926, doi:10.1080/00224499.2015.1062840.

6. Emily A. Vogels and Monica Anderson, "Dating and Relationships in the Digital Age," Pew Research Center, May 8, 2020, https://www.pewresearch.org/internet/2020/05/08/dating-and-relationships-in-the-digital-age; Jason Dibble, J. Banas, and Michelle Drouin, "Communication and Romantic Alternatives: Keeping Ex-Partners on the Back Burner" (unpublished manuscript, December 2020).

7. Brandon T. McDaniel, Michelle Drouin, and Jaclyn D. Cravens, "Do You Have Anything to Hide? Infidelity-Related Behaviors on Social Media

Sites and Marital Satisfaction," *Computers in Human Behavior* 66 (January 2017): 88–95, https://doi.org/10.1016/j.chb.2016.09.031.

8. James Sexton, "Divorce Lawyer: Facebook Is a Cheating Machine," *Time*, March 26, 2018, https://time.com/5208108/facebook-cheating -infidelity-divorce.

9. "Marriage and Unions," Department of Economic and Social Affairs, United Nations, https://www.un.org/en/development/desa/population /theme/marriage-unions/index.asp.

10. "Changing Patterns of Marriage and Unions across the World," Department of Economic and Social Affairs, United Nations, https://www.un .org/en/development/desa/population/publications/factsheets/index.asp

11. Natalie Nitsche and Sarah R. Hayford, "Preferences, Partners, and Parenthood: Linking Early Fertility Desires, Marriage Timing, and Achieved Fertility," *Demography* 57, no. 6 (December 2020): 1975–2001, doi:10.1007 /s13524-020-00927-y.

12. John Casterline and Han Siqi, "Unrealized Fertility: Fertility Desires at the End of the Reproductive Career," *Demographic Research* 36, no. 14 (January 2017): 427–454, doi:10.4054/DemRes.2017.36.14.

13. Tomáš Sobotka and Éva Beaujouan, "Two Is Best? The Persistence of a Two-Child Family Ideal in Europe," *Population and Development Review* 40, no. 3 (September 2014): 391–419, https://doi.org/10.1111/j.1728 -4457.2014.00691.x.

14. Gretchen Livingston, "They're Waiting Longer, but U.S. Women Today More Likely to Have Children than a Decade Ago," Pew Research Center, https://www.pewresearch.org/social-trends/2018/01/18/theyre-waiting -longer-but-u-s-women-today-more-likely-to-have-children-than-a-decade -ago.

15. "Infertility," National Center for Health Statistics, Centers for Disease Control and Prevention, https://www.cdc.gov/nchs/fastats/infertility.htm; "Global Prevalence of Infertility, Infecundity and Childlessness," Sexual and Reproductive Health, World Health Organization, https://www.who .int/reproductivehealth/topics/infertility/burden/en.

16. Jennifer Glass, Robin W. Simon, and Matthew A. Andersson, "Parenthood and Happiness: Effects of Work-Family Reconciliation Policies in 22

OECD Countries," *American Journal of Sociology* 122, no. 3 (November 2016): 886–929, doi:10.1086/688892.

17. Debra Umberson, "Gender, Marital Status, and the Social Control of Health Behavior," *Social Science and Medicine* 34, no. 8 (April 1992): 907–917, https://doi.org/10.1016/0277-9536(92)90259-S.

18. Linda Waite and Maggie Gallagher, *The Case for Marriage: Why Married People Are Happier, Healthier, and Better Off Financially* (New York: Doubleday, 2001).

19. Jody VanLaningham, David R. Johnson, and Paul Amato, "Marital Happiness, Marital Duration and the U-Shaped Curve: Evidence from a Five-Wave Panel Study," *Social Forces* 79, no. 4 (June 2001): 1313–1341, doi:10.1353/sof.2001.0055.

20. Thibault and Kelley, *Social Psychology of Groups*.

21. Robert J. Sternberg, "Duplex Theory of Love: Triangular Theory of Love and Theory of Love as a Story," http://www.robertjsternberg.com/love.

22. Cyrille Feybesse and Elaine Hatfield, "Passionate Love," in *The New Psychology of Love*, 2nd ed., ed. Robert J. Sternberg and Karin Sternberg (Cambridge: Cambridge University Press, 2019), 183–207.

23. Robert J. Sternberg, "A Triangular Theory of Love," *Psychological Review* 93, no. 2 (1986): 119–135, https://doi.org/10.1037/0033-295X.93.2.119.

24. Piotr Sorokowski, Agnieszka Sorokowska, Maciej Karwowski, Agata Groyecka, Toivo Aavik, Grace Akello, Charlotte Alm, et al., "Universality of the Triangular Theory of Love: Adaptation and Psychometric Properties of the Triangular Love Scale in 25 Countries," *Journal of Sex Research* 58, no. 1 (August 2020): 106–115, doi:10.1080/00224499.2020.1787318.

25. Gian C. Gonzaga, "Romantic Love and Sexual Desire in Close Relationships," *Emotion* 6, no. 2 (2006): 163–179: https://doi.org/10.1037/1528-3542.6.2.163; Elaine Hatfield and Richard L. Rapson, "Passionate Love, Sexual Desire, and Mate Selection: Cross-Cultural and Historical Perspectives," in *Close Relationships: Functions, Forms and Processes*, ed. Patricia Noller and Judeith A. Feeney (Oxfordshire, UK: Psychology Press, 2006), 227–243.

26. James K. McNulty, Carolyn A. Wenner, and Terri D. Fisher, "Longitudinal Associations among Relationship Satisfaction, Sexual Satisfaction, and

Frequency of Sex in Early Marriage," *Archives of Sexual Behavior* 45, no. 1 (December 2014): 85–97, doi:10.1007/s10508-014-0444-6.

27. Trond Viggo Grøntvedt, Leif Edward Ottesen Kennair, and Mons Bendixen, "How Intercourse Frequency Is Affected by Relationship Length, Relationship Quality, and Sexual Strategies Using Couple Data," *Evolutionary Behavioral Sciences* 14, no. 2 (April 2019): 147–159, doi:10.1037/ebs0000173.

28. E. Sandra Byers and Larry Heinlein, "Predicting Initiations and Refusals of Sexual Activities in Married and Cohabiting Heterosexual Couples," *Journal of Sex Research* 26, no. 2 (1989): 210–231, http://dx.doi.org/10.1080/00224498909551507.

29. Amelia Karraker, John DeLamater, and Christine R. Schwartz, "Sexual Frequency Decline from Midlife to Later Life," *Journals of Gerontology: Series B* 66B, no. 4 (July 2011): 502–512, doi:10.1093/geronb/gbr058.

30. Donald G. Dutton and Arthur P. Aron, "Some Evidence for Heightened Sexual Attraction under Conditions of High Anxiety," *Journal of Personality and Social Psychology* 30, no. 4 (November 1974): 510–517, doi:10.1037/h0037031.

31. Vaughn Call, Susan Sprecher, and Pepper Schwartz, "The Incidence and Frequency of Marital Sex in a National Sample," *Journal of Marriage and the Family* 57, no. 3 (August 1995): 639–652, http://dx.doi.org/10.2307/353919.

32. Colette Hickman-Evans, Jesse P. Higgins, Ty Aller, Joy Chavez, and Kathy W. Piercy, "Newlywed Couple Leisure: Couple Identity Formation through Leisure Time," *Marriage and Family Review* 54, no. 2 (March 2017): 105–127, doi:10.1080/01494929.2017.1297756.

33. Michelle Drouin and Brandon T. McDaniel, "Technology Use during Couples' Bedtime Routines, Bedtime Satisfaction, and Associations with Individual and Relational Well-being," *Journal of Social and Personal Relationships*, February 10, 2021, https://doi.org/10.1177/0265407521991925.

34. George Loewenstein, Tamar Krishnamurti, Jessica Kopsic, and Daniel McDonald, "Does Increased Sexual Frequency Enhance Happiness?," *Journal of Economic Behavior and Organization* 116 (August 2015): 206–218.

35. Bernie Zilbergeld and C. R. Ellison, "Desire Discrepancies and Arousal Problems in Sex Therapy," in *Principles and Practice of Sex Therapy*, ed. Sandra R. Leiblum and Lawrence A. Pervin (New York: Guilford Press, 1980), 65–106.

36. Marieke Dewitte, Joana Carvalho, Giovanni Corona, Erika Limoncin, Patrícia M. Pascoal, Yacov Reisman, and Aleksandar Štulhofer, "Sexual Desire Discrepancy: A Position Statement of the European Society for Sexual Medicine," *Sexual Medicine* 8, no. 2 (June 2020): 121–131, doi:10.1016/j.esxm.2020.02.008.

37. Giovanni Corona, Andrea M. Isidori, Antonio Aversa, Arthur L. Burnett, and Mario Maggi, "Endocrinologic Control of Men's Sexual Desire and Arousal/Erection," *Journal of Sexual Medicine* 13, no. 3 (March 2016): 317–337, doi:10.1016/j.jsxm.2016.01.007.

38. Byers and Heinlein, "Predicting Initiations and Refusals of Sexual Activities."

39. Megan E. McCool, Andrea Zuelke, Melissa A. Theurich, Helge Knuettel, Cristian Ricci, and Christian Apfelbacher, "Prevalence of Female Sexual Dysfunction among Premenopausal Women: A Systematic Review and Meta-analysis of Observational Studies," *Sexual Medicine Reviews* 4, no. 3 (July 2016): 197–212, doi:10.1016/j.sxmr.2016.03.002.

40. Roy F. Baumeister, Kathleen R. Catanese, and Kathleen D. Vohs, "Is There a Gender Difference in Strength of Sex Drive? Theoretical Views, Conceptual Distinctions, and a Review of Relevant Evidence," *Personality and Social Psychology Review* 5, no. 3 (August 2001): 242–273, doi:10.1207/S15327957PSPR0503_5.

41. R. C. Rosen, "Prevalence and Risk Factors of Sexual Dysfunction in Men and Women," *Current Psychiatry Reports* 2, no. 3 (June 2000): 189–195, doi:10.1007/s11920-996-0006-2.

42. Gurit E. Birnbaum, "The Fragile Spell of Desire: A Functional Perspective on Changes in Sexual Desire across Relationship Development," *Personality and Social Psychology Review* 22, no. 2 (May 2018): 101–127, doi:10.1177/1088868317715350.

43. Nicholas M. Grebe, Steven W. Gangestad, Christine E. Garver-Apgar, and Randy Thornhill, "Women's Luteal-Phase Sexual Proceptivity and the

Functions of Extended Sexuality," *Psychological Science* 24, no. 10 (August 2013): 2106–2110, https://doi.org/10.1177/0956797613485965.

44. Elisa Ventura-Aquino, Alonso Fernández-Guasti, and Raúl G Paredes, "Hormones and the Coolidge Effect," *Molecular and Cellular Endocrinology* 467 (May 2018): 42–48, doi:10.1016/j.mce.2017.09.01; Dennis F. Fiorino, Ariane Coury, and Anthony G. Phillips, "Dynamic Changes in Nucleus Accumbens Dopamine Efflux during the Coolidge Effect in Male Rats," *Journal of Neuroscience* 17, no. 12 (June 1997): 4849–4855, https://doi.org/10.1523/JNEUROSCI.17-12-04849.1997.

45. Helen E. Fisher, "Lust, Attraction, and Attachment in Mammalian Reproduction," *Human Nature* 9, no. 1 (1998): 23–52, doi:10.1007/s12110-998-1010-5.

46. Ogi Ogas and Sai Gaddam, *A Billion Wicked Thoughts: What the Internet Tells Us about Sex and Relationships* (New York: Plume, 2012).

47. "Robot Companion," Robot Companion, https://www.robotcompanion.ai.

48. Matthew Dunn, "Human-esque Sex Robots, Connected Toys and VR Are the Future of the Adult Industry," news.com.au, July 30, 2017, https://www.news.com.au/technology/innovation/design/humanesque-sex-robots-connected-toys-and-vr-are-the-future-of-the-adult-industry/news-story/f2b7e8eb091aea846bd0c1cf38c4488b.

CHAPTER 7

1. David M. Buss and David P. Schmitt, "Sexual Strategies Theory: An Evolutionary Perspective on Human Mating," *Psychological Review* 100, no. 2 (1993): 204–232, https://doi.org/10.1037/0033-295X.100.2.204.

2. Satoshi Kanazawa and Mary C. Still, "Is There Really a Beauty Premium or an Ugliness Penalty on Earnings?," *Journal of Business and Psychology* 33, no. 2 (April 2018): 249–262, doi:10.1007/s10869-017-9489-6.

3. Laura Wood, "Anti-Aging Products Industry Projected to Be Worth $83.2 Billion by 2027—Key Trends, Opportunities and Players," Research and Markets, Intrado Globe Newswire, July 24, 2020, https://www.globenewswire.com/news-release/2020/07/24/2067180/0/en/Anti-Aging-Products-Industry-Projected-to-be-Worth-83-2-Billion-by-2027-Key-Trends-Opportunities-and-Players.html.

4.   Yong Liu, Janet B. Croft, Anne G. Wheaton, Dafna Kanny, Timothy J. Cunningham, Hua Lu, Stephen Onufrak, et al., "Clustering of Five Health-Related Behaviors for Chronic Disease Prevention among Adults, United States, 2013," *Preventing Chronic Disease* 13 (May 2016): 160054, http://dx.doi.org/10.5888/pcd13.160054.

5.   Eric N. Reither, Robert M. Hauser, and Yang Yang, "Do Birth Cohorts Matter? Age-Period-Cohort Analyses of the Obesity Epidemic in the United States," *Social Science and Medicine* 69, no. 10 (November 2009): 1439–1448, doi:10.1016/j.socscimed.2009.08.040.

6.   S. Jay Olshansky, Douglas J. Passaro, Ronald C. Hershow, Jennifer Layden, Bruce A. Carnes, Jacob Brody, Leonard Hayflick, et al., "A Potential Decline in Life Expectancy in the United States in the 21st Century," *New England Journal of Medicine* 352, no. 11 (March 2005): 1138–1145, doi:10.1056/NEJMsr043743; Steven H. Woolf and Heidi Schoomaker, "Life Expectancy and Mortality Rates in the United States, 1959–2017," *Journal of the American Medical Association* 322, no. 20 (November 2019): 1996–2016, doi:10.1001/jama.2019.16932.

7.   National Academies of Sciences, Engineering, and Medicine, *Social Isolation and Loneliness in Older Adults: Opportunities for the Health Care System*, 2020, https://www.nap.edu/catalog/25663/social-isolation-and-loneliness-in-older-adults-opportunities-for-the.

8.   "Loneliness and Social Isolation Linked to Serious Health Conditions," Alzheimer's Disease and Healthy Aging, Centers for Disease Control and Prevention, https://www.cdc.gov/aging/publications/features/lonely-older-adults.html.

9.   James E. Lubben and M. W. Gironda, "Centrality of Social Ties to the Health and Well-being of Older Adults," in *Social Work and Health Care in an Aging World*, ed. Barbara Berkman and Linda Harooytan (New York: Springer, 2003), 319–350.

10.   James S. House, "Social Isolation Kills, but How and Why?," *Psychosomatic Medicine* 63, no. 2 (March 2001): 273–274, doi:10.1097/00006842-200103000-00011.

11.   National Academies of Sciences, Engineering, and Medicine, *Social Isolation and Loneliness in Older Adults*.

12. House, "Social Isolation Kills."

13. Nicholas G. Castle, John Engberg, and Aiju Men, "Nursing Home Staff Turnover: Impact on Nursing Home Compare Quality Measures," *Gerontologist* 47, no. 5 (October 2007): 650–661, doi:10.1093/geront/47.5.650.

14. Corinna Vossius, Geir Selbæk, Jurate Šaltytė Benth, and Sverre Bergh, "Mortality in Nursing Home Residents: A Longitudinal Study over Three Years," *PLoS One* 13, no. 9 (September 2018): 1–11, doi:10.1371/journal.pone.0203480.

15. Jennifer Casarella, "Dealing with Chronic Illnesses and Depression," WebMD, September 27, 2020, https://www.webmd.com/depression/guide/chronic-illnesses-depression#1.

16. Thomas F. Hack and Lesley F. Degner, "Coping Responses following Breast Cancer Diagnosis Predict Psychological Adjustment Three Years Later," *Psycho-Oncology* 13, no. 4 (June 2003): 235–247, doi:10.1002/pon.739.

17. "U.S. Breast Cancer Statistics," BreastCancer.org, last modified February 4, 2021, https://www.breastcancer.org/symptoms/understand_bc/statistics.

18. Veena Shukla Mishra and Dhananjaya Saranath, "Association between Demographic Features and Perceived Social Support in the Mental Adjustment to Breast Cancer," *Psycho-Oncology* 28, no. 3 (January 2019): 629–634, doi:10.1002/pon.5001.

19. Jeana H. Frost and Michael P Massagli, "Social Uses of Personal Health Information within PatientsLikeMe, an Online Patient Community: What Can Happen When Patients Have Access to One Another's Data," *Journal of Medical Internet Research* 10, no. 3 (May 2008): e15, doi:10.2196/jmir.1053.

20. Paul Wicks, Michael Massagli, Jeana Frost, Catherine Brownstein, Sally Okun, Timothy Vaughan, Richard Bradley, and James Heywood, "Sharing Health Data for Better Outcomes on Patients like Me," *Journal of Medical Internet Research* 12 no. 2 (June 2010): 117–128. doi:10.2196/jmir.1549.

21. Iroju Olaronke, Abimbola Soriyan, Ishaya Gambo, and J. Olaleke, "Interoperability in Healthcare: Benefits, Challenges and Resolutions," *International Journal of Innovation and Applied Studies* 3, no. 1 (April 2013): 262–270.

22. The Clinical and Business Imperative for Healthcare Organisations, "Strategic Interoperability in Germany, Spain and the UK," 2014, https://www.digitalhealthnews.eu/download/white-papers/3947-strategic-interoperability-in-germany-spain-and-the-uk-the-clinical-and-business-imperative-for-healthcare-organisations.

23. Hannah Crouch, "Greater Manchester, Wessex and One London Selected as LHCREs," Digital Health, May 23, 2018, https://www.digitalhealth.net/2018/05/greater-manchester-wessex-and-one-london-lhcre.

24. "Enterprise Imaging Solutions," Watson Health, IBM, https://www.ibm.com/watson-health/solutions/enterprise-imaging.

25. Owen Hughes, "Tech Giants Make Interoperability Pledge for US Health Data," Digital Health, August 17, 2018, https://www.digitalhealth.net/2018/08/tech-giants-make-interoperability-pledge-for-us-health-data.

26. "AWS Data Exchange," AWS Marketplace, Amazon, https://aws.amazon.com/data-exchange.

27. "Coronavirus (COVID-19) Data Hub," Tableau, AWS Marketplace, Amazon, https://aws.amazon.com/marketplace/pp/prodview-a5mqede4xd4c4?qid=1609113770043&sr=0-1&ref_=srh_res_product_title.

28. "Standards Development Organizations," Office of the National Coordinator for Health Technology, HealthIT.gov, https://www.healthit.gov/playbook/sdo-education/chapter-2.

29. "Interoperability in Healthcare," Healthcare Information and Management Systems Society, https://www.himss.org/resources/interoperability-healthcare.

30. "HHS Finalizes Historic Rules to Provide Patients More Control of Their Health Data," US Department of Health and Human Services, March 9, 2020, https://www.hhs.gov/about/news/2020/03/09/hhs-finalizes-historic-rules-to-provide-patients-more-control-of-their-health-data.html.

Index

Amazon, 84, 97, 134, 218, 221
  AWS Data Exchange, 221
American Academy of Pediatrics, 23,
    70, 73, 75, 77, 88
American National Standards
    Institute, 222
American Psychological Association
    (APA), 108
Analysis paralysis, 22, 147, 151, 164
Anna Karenina, 152
Antonucci, Toni, 11
Apple, 75, 218, 220
  Health Records application, 220
  research application, 220
Aron, Arthur, 184
Artificial general intelligence (AGI),
    3
Ashleymadison.com, 172
Assortative mating, 140, 143
Attraction, 36, 96, 181, 184–185,
    193–195, 199–200

Baby Einstein, 70
Bachelor, The, 184

Back burners, 163–164, 171, 173,
    196
Bambi, 52
Bandura, Albert, 131
BDSM, 115
Beauty premium, 206
Bell, Alexander Graham, 76
Bem, Daryl, 154–156
Bentham, Jeremy, 113
Bentham, Samuel, 113
Biederman, Irving, 125
Blass, Elliott, 59
Bodily fingerprint, 156
Brady, Tom, 64, 67–68, 85
British Household Panel Study,
    141
Bumble, 108, 138
Buss, David, 142, 147, 205

Cambridge Analytica, 115
Castaway, 110
Centers for Disease Control and
    Prevention (CDC), 206–208,
    221

Clemens, Samuel, 69, 89–90, 93, 109–110
Clooney, George, 200
Cohen, Sheldon, 46, 122–123
Cohen, Stanley, 116
Collins, Judy, 65
Commitment, 181–185, 193, 196–197, 231
Common Sense Media, 72, 77–78, 88
Convoy model of social relations, 11
Coolidge effect, 194
Cost of Loneliness Project, 107
Costanza, George, 200
COVID-19 pandemic, 29–33, 37, 53, 97, 221
Coyle, John, 169
Crumple zone, 210
Cuddling
  cuddle parties, 51, 56
  Cuddlist, 50–51

Dater's potential decisions
  determining relationship structure or duration, 153–154
  initial search, 153
  steps to help determine there is a connection, 153
Denial, 116
Denworth, Lydia, 94
Destroy after reading, 129–130, 135, 163
Dibble, Jayson, 163, 172
Didion, Joan, 9
Diff'rent Strokes, 79
Disney, 4, 69–70, 148
Dopamine, 10, 14, 45, 51, 105–106, 125, 134, 185, 194

Dostoyevsky, Fyodor, 120
Dunbar, Robert, 104
Dutton, Donald, 184

Ekman, Paul, 4
Erikson, Erik, 33–36, 93
Etwaru, Richie, 222
European Society for Sexual Medicine (ESSM), 189–190

Family Romance, 108
Federal Trade Commission, 70
Fight Club, 170–171
Ford, Henry, 76
Forecasting errors, 100
Fortnite, 71
Franklin, Benjamin, 76
Friendship, 89–109
  friends are like money, 111–112
  how to survive, 109–111

Gallup World Poll, 9, 107
Gates, Bill, 75
General Data Protection Regulation, 115
General Social Survey, 10, 13–14, 19
Generation Me, 9
Ghosting, 132–135, 165
Girl Scouts, 90–92, 110
Glueck, Eleanor, 81
Glueck, Sheldon, 81
Glueck Study, 81
Goal-setting theory of motivation, 18
Grant, W. T., 81
Grant Study, 81
Growing old, how to survive, 225–228
Guinazzo, Madelon, 51

Habituation, 52, 180, 193–194, 200

Hancock, Jeffrey, 125

Hanson, David, 4

Happiness chemicals, 39

Harlow, Harry, 64

Head Start, 69, 71

Healthcare Information and
  Management Systems Society,
  219–220

Hemingway, Ernest, 170

Hepburn, Audrey, 4

Heterogamy, 140–141

Hinge, 35

Homogamy, 140–141, 143

House, James, 209–210

Hu-manity.co, 222

Hypergamy, 144

IBM
  iConnect, 220
  Watson Health, 220
iGen, 9

Illness, 212–217, 226

Infovore, 126

Interdependence theory, 162

Internal working models, 67

International Organization for
  Standardization, 222

Intimacy famine, 4, 10, 14, 17, 45,
  175

Investment model, 162, 172

I, Robot, 111

James, Aaron, 53

Jingle-jangle problem, 103

Johnson, Lyndon, 69

Jones, Benjamin, 37

Just world phenomenon, 109

Kahn, Robert, 11

Kelley, Harold, 162

Khan Academy, 83

Kim, Jonny, 145

Klein, Nadav, 158–159

Kondo, Marie, 128

Kross, Ethan, 103

Life expectancy, 13, 24, 208

Loewenstein, George, 14, 16–17,
  189

Loneliness, 107–108, 121, 179, 208–
  209, 230

Lubben Social Network Scale, 208

Luo, Mufan, 125

Marasmus, 63

Mare, Robert, 141

Marriage
  benefits of, 179
  how to survive, 197–199
  sexless marriage, 17

Maslow, Abraham, 42, 131

Maslow's Hierarchy of Needs, 131

Maximizers, 152

Mayer, John, 96

McDaniel, Brandon, 118, 172, 186

McGrath, Maureen, 17

McNulty, James, 191

Mesolimbic pathway, 105, 194

Miller, Dan, 163, 172

Mischel, Walter, 154, 157

Mister Rogers, 68–70, 86

Monchhichis, 90, 92, 110

National Geographic, 125, 211

Need for closure, 132

Needle in a haystack, 152, 164

Netflix, 71, 87, 224
Nietzsche, Friedrich, 7, 9

O'Brien, Ed, 158–159
Office, The, 43
Ogg, David, 62
Ogg, Katie, 62
Ong, David, 143
Online dating, 108, 138–139, 148–149, 152, 161, 165
Oxford COVID-19 Government Response Tracker, 31

Pandemic losses, 54
Pandemic puppies, 38, 55
Paradox of Choice, 22
Parks and Recreation, 43
Passion, 166, 181–185, 193
Patients-LikeMe (PLM), 216–218, 223
Persistence, 130, 162
Piaget, Jean, 72, 138
Plenty of fish in the sea, 147, 161, 163–165, 168
Pollyanna principle, 129
Positive affect bias, 174
Psi phenomena, 154–155, 167
Psychological Care of Infant and Child, 61
Psychological reactance, 16, 99

Quaas, Johanna, 206
Quality of alternatives, 162

RAND Health and Retirement study, 215
Reinforcers, types of
  primary reinforcers, 105–106
  secondary reinforcers, 105

RentAFriend, 108
Replika, 6–9
Responsiveness, 53, 73, 118–119, 190
Reward pathway, 105, 185
Romele, Alberto, 116
Rose, Lucy, 107
Rumination, 129
Rusbult, Caryl, 162

Sachs, Jeffrey, 10
Sandlot, The, 74
Schmitt, David, 205
Schwartz, Barry, 22, 151
Schwartz, Christine, 141, 183
Self-perception theory, 156
Sesame Street, 65, 69, 71
Sexting, 173, 198
Sexual desire, 17, 183, 189–193, 196, 199–200
Sexual desire discrepancy, 17, 189–190
Shackelford, Todd, 142, 147
Signal detection theory, 152
Skin-to-skin contact, 60–62, 85
Snowden, Edward, 114
Social brain hypothesis, 104
Social capital, 10, 111, 123
Social Dilemma, The, 120
Social economizing, 97–98, 167–168
Social exchange theory, 180
Social isolation, 37, 64, 100, 110, 208–209
Social Network Index (SNI), 121–122
Social support, 123–126, 134, 216–217
Sociosexual orientation, 183

Sociotechnical panopticon, 113, 133

Sophia the Robot, 1–6, 24–25,
224–227

Spitz, René, 63

Strangers
talking to, 95–99
touching, 47

Support, categories of
appraisal, 124
emotional, 124
informational, 124
instrumental, 124

Symbolic interactionism, 119

Technoference or phubbing,
118–120

TEDx, 17, 169–171

Theory of mind, 127–128

Thibaut, John, 162

TikTok videos, 80, 87

Tinder, 20, 138, 144, 150

Triangular theory of love, 181

Twenge, Jean, 9–10, 20, 120

U.S. National Longitudinal Survey of
Youth, 177

Uncertainty reduction theory, 132

United Nations, 175–176

Unrealized fertility, 177–178

Unrealized intimacy, 186–188, 197

Vaillant, George, 81

Vessel, Edward, 125

Vygotsky, Lev, 72

Wallace, David Foster, 8

Watson, John, 61

Westworld, 2, 4

Wizard of Oz, The, 53, 69

World Economic Outlook, 44

World Happiness Report, 9–10

World Health Organization, 73,
75, 77

World Marriage Data, 175

Year of Magical Thinking, The, 9

Zoom fatigue, 44